The Evolution of Complexity
by Means of Natural Selection

BOOKS BY JOHN TYLER BONNER

Morphogenesis

Cells and Societies

The Evolution of Development

The Cellular Slime Molds

The Ideas of Biology

Size and Cycle

The Scale of Nature

On Development

The Evolution of Culture in Animals

On Size and Life (with T. A. McMahon)

The Evolution of Complexity

John Tyler Bonner

The Evolution of Complexity

by Means of Natural Selection

PRINCETON UNIVERSITY PRESS
PRINCETON, NEW JERSEY

Published by Princeton University Press, 41 William Street,
Princeton, New Jersey 08540
In the United Kingdom: Princeton University Press,
Guildford, Surrey

Library of Congress Cataloging-in-Publication Data

Bonner, John Tyler.
The evolution of complexity by means of
natural selection.

Bibliography: p.
Includes index.
1. Evolution. 2. Natural selection. I. Title.
II. Title: Evolution of complexity.
QH371.B65 1988 575 87-38490
ISBN 0-691-08493-9 (alk. paper)
ISBN 0-691-08494-7 (pbk.)

This book has been composed in Linotron Times Roman

Clothbound editions of Princeton University Press books
are printed on acid-free paper, and binding materials are
chosen for strength and durability. Paperbacks, although
satisfactory for personal collections, are not usually
suitable for library rebinding

Printed in the United States of America by
Princeton University Press,
Princeton, New Jersey

Contents

Preface

THE MAIN ideas in this book have been fermenting inside me for many years. My first resolution to do something about putting them down came from the prodding of a Dahlem conference on evolution and development in 1982. It was a gathering of evolutionary biologists (including paleontologists), developmental biologists, and molecular biologists who brought the views of their separate disciplines to bear on the relation of development to the study of evolution and vice versa. For many of us it was a useful and stimulating experience; but as I edited the volume and carefully scrutinized each contribution, I was made acutely aware that it was not an integrated book, but a collection of interesting, yet divergent perspectives. It was then that I began to contemplate the idea of bringing the subject together and presenting it from one point of view, but I did not want simply to duplicate what others had done so effectively before me, especially G. R. de Beer (1958), S. J. Gould (1977), and R. A. Raff and T. C. Kaufman (1983). It seemed to me there was more to be said and that the time was right for me to put my thoughts in order and on paper.

This is not confined solely to the relation of development to evolution, but it has a broader perspective. Besides development it involves the relation of genetics, community structure, and animal behavior to evolution because they are all so closely intertwined. My examination of this network of subjects brought me ultimately to an old question that goes far back into the history of biology; it is the question of progress. How is it that an egg turns into an elaborate adult; how is it that a bacterium, given many millions of years, could have evolved into an elephant. During the major part of the last century and well into this one, many biologists believed that such advances were accomplished by a progressive force, unspecified and mystical, the only evidence of which came from the extraordinary results which are there for all to see. A more mechanistic approach to this evolution of complexity was adopted by Charles Darwin, who conceived the idea that the changes which could be called progress had their explanation in natural selection. Yet despite the great success of his writings, the number of biologists favoring his view of the role of selection was insignificant until well into this century. The progressive ''scale of nature'' is a subject that has fascinated all biologists from before Aristotle to this day, and I am no exception.

I am not a philosopher but a biologist, and it is the purpose of this book to seek biological answers to some important biological problems. My interest began some years ago with trying to understand the mechanism of development and I chose to work on cellular slime molds, a curious group of primitive multicellular organisms that seemed to have features especially adapted to the study of what Wilhelm Roux called the mechanics of development. In those days I also, as a counterweight to the peculiarities of slime molds, worked, often in collaboration with gifted students, on the development of ciliate protozoa, hydroids, filamentous fungi, and colonial algae. From this diversity of points of view I inevitably became increasingly concerned with the question, why are there so many different kinds of development; why did they evolve? All roads in biology lead to evolution and the principle of natural selection; it would have been impossible to miss the turn.

The first step was the appreciation that there are many levels of biological organization and that, for instance, there are close parallels between the organization of cells within a multicellular individual and that of a group of individuals in an animal society. The parallels are no mere analogy, for the same forces of natural selection act at both levels. Among other things there is a difference in size between the levels, and this is indeed important, a matter I have examined in various ways. The problem of differences in complexity was temporarily put in abeyance because it seemed less tractable, and what to do about it was not immediately obvious.

Instead I pursued other themes, especially that of animal behavior. Again there are so many parallels between behavior (and its development) and the development of bodily structure that I thought of behavior as an extension of development, one that involved a great increase in plasticity, which in itself is often selectively advantageous. One of the windfalls of flexible behavior is an increase in the ability of an animal to learn, and even teach, and from this arose culture, of which we, as human beings, are so conscious. The important point, however, is that there is an enormous difference between the mode of inheritance of flexible behavior and of structure. The latter is directly controlled by genes, while the former is passed from one individual to another by nongenetic or behavioral transmission. These are two kinds of inheritance which operate by totally different mechanisms, but dovetail with one another and are both under the direct influence of selection. As I shall discuss in detail in this book, both are ultimately controlled by genes, but the connection between flexible behavior and the genome is remote; what is inherited is a capacity to learn and invent and instruct. This capacity to vary in a way which is not directly controlled by genes is also characteristic of cell differentiation in a multicellular organism, for our muscle cells and our nerve cells and our liver cells are all genetically identical; their differences arose by gene-directed cytoplasmic actions and reactions. Therefore both behavior and cell differentiation are to varying degrees removed from the immediate gene instructions.

The rounding of my education came in the 1960s when the new vistas of population biology opened up to me. This was largely due to my association with my good friends and colleagues, Robert MacArthur and Henry Horn, who showed me their clear and revealing insights into the world of ecology. A different kind of complexity existed at this highest level, and what concerned me was its relevance to lower levels of organization.

All of these varied interests have now been put to bear on the problem of the evolution of complexity. To state it in the simplest terms: Why has there been an evolution from the primitive bacteria of billions of years ago to the large and complex organisms of today? The answer to the question of "why" is straightforward: it is natural selection. But my interest lies at a deeper plane and I want to know how selection has achieved this progress at each level—that of the cell, the multicellular organism, the animal society, and the ecological community.

If I am to talk about such levels of complexity I will automatically drift into the province of the holists and the general-systems theorists who have made a philosophical study of levels of organization and how they relate to one another. Unfortunately, despite repeated efforts, I have not been able to find their ideas useful. Much of the impetus for them seems to be an anger at reductionism, which they consider narrow and binding and, to some, antithetic to dialectical materialism. What is utterly baffling to me is why one cannot be a reductionist and a holist at the same time. It is the reductionist who supplies particularly satisfying explanations, and it is the holist who takes the grand view and can show what questions need answering. This is the way science has worked in the past, and certainly all the great advances in biology over the last two centuries have been made by reductionist microexplanations, and this includes the explanation of evolution by natural selection. Yet reductionism alone is not enough; one must look at the larger picture too. For instance, today molecular biology, the ultimate in biological reductionism, would be lost without the cell biologist, the developmental biologist, the geneticist, the neurobiologist, and the evolutionist to guide the way.

The discussion of "levels" that follows in this book will therefore seem far removed from the hierarchy theory of L. von Bertalanffy, H. H. Pattee, and many others. My concern is how the levels arose, how they are constructed during development, how they are genetically controlled, and how they relate to one another. These are biological concerns that can all be answered by examining them in terms of natural selection. Since this is so, the explanation sought will be of the micro or reductionist sort, for each component of each level will be considered in terms of the selective forces that could account for its first appearance and maintenance during the course of evolution. As has so often been pointed out in the past, a good explanation is one that gives some inner satisfaction, and therefore it is a very human desire to seek good explanations. But not all human beings are the same, and for some a religious or mystical explanation is the most satisfying. This is not so for me: the more rational and materialistic the explanation, the better I like it.

This book begins with a brief and elementary chapter on evolution. It is a summary of how we think of evolution today. The next three chapters are concerned with the evolution of size change, especially size increase, a subject that is essential in the consideration of complexity. First there is a discussion of the paleontological evidence (Chapter 2) and then one on the size distribution of animals and plants in nature (Chapter 3). Since all organisms begin each generation from a single cell, a fertilized egg, or an asexual spore, an increase in size has meant changes in the engineering of growth and development (Chapter 4). The remainder of the book fixes directly on the evolution of complexity, beginning with the complexity of ecological communities discussed primarily in terms of species diversity (Chapter 5). The discussion of complexity in development (Chapter 6) is a particularly important part of my thesis since it examines how genes can control developmental events that are far removed from the immediate gene products; there is a complex web of actions and reactions that are started, and only remotely controlled, by the genes. Behavior is even further removed from the genes and largely for this reason becomes the pinnacle of biological plasticity (Chapter 7). From all this it is possible to propose three general insights (Chapter 8) which shed light on the evolution of complexity.

THE first draft of this book was read by my good friend and colleague, Henry Horn. He is the ideal critic who tells one what is wrong with such precision and clarity that one is enormously helped in making improvements. My debt to him for help with this book, and many of my writings in the past, is very large. The second draft received two treatments. It was read by all the students and auditors in my graduate course at Princeton (in the fall of 1985). Their comments on the text, and especially on my lectures based on the text, were invaluable to me. Furthermore, the students gave papers at the end of the course on relevant topics of their own choosing which were rewarding to us all, and they will see signs of their contributions in the third draft. I wish to thank the following undergraduates: Stephen Barr, Jill Kraft, Samuel Moskowitz, John Scott, Letitia Volpp; the following graduate students: Joshua Ginsberg, Graham Head, Anna Marie Lyles, Margaret McFarland, Nancy Pratt, William Zerges; and the following colleagues: Jane Brockman, Andrew Dobson, Henry Horn, William Jacobs, Jon Seger; and my research assistant Hannah Suthers. This draft was also sent to three outside reviewers: Donna Bozzone, Leo Buss, and Todd Newberry. Their generosity in going over the text with a broad-toothed comb (for ideas) and a fine-toothed comb (for infelicitous phrasing) is without limit, and I am correspondingly in their debt. They were, like Henry Horn, the best of critics, and they, too, often gave me the uneasy feeling that they understood my ideas better than I did. I am also grateful to Alexander Bearn, Robert Brandon, Ralph Greenspan, Robert May, and George Oster for their helpful comments.

The bulk of this book was written while I was on leave during the academic year 1984–85. The fall and the spring were spent in the Zoology Department at the University of Edinburgh, which has become a second biological home for me. The stay was so very pleasant largely due to the many kindnesses of Murdoch and Rowy Mitchison. The winter months were spent in a small town in Mallorca as guests of my brother Tony and his wife Eve. Tony interrupted his work on medieval philosophy and Eve her work to ensure that conditions for my writing were optimal, and they succeeded beyond my greatest hopes. These are all debts that cannot ever be adequately repaid. My wife Ruth accompanied and helped me through all these travels and has been endlessly patient in smoothing my rough prose.

I would like to thank Elizabeth Begley from Edinburgh for typing the first draft, Linda Karanewsky for putting the second draft on the computer, and Mary Leksa for coping so cheerfully and effectively with the massive, complicated changes of the third draft.

Finally I would like to give special thanks to the Commonwealth Fund Book Program and its director, Lewis Thomas, for their support during the course of the writing of this book.

August 1987 J.T.B.
Margaree Harbour
Cape Breton
Nova Scotia

The Evolution of Complexity
by Means of Natural Selection

Chapter 1

A Brief Summary of Darwinian Evolution, along with an Indication of the Purpose of the Book

Time • Natural selection • Development • Ecology • Behavior • Genetics, development, ecology, behavior, and evolution

BECAUSE evolution is such a big subject, it is almost invariably true that any current treatment will examine only one or a few of its facets, but here I will bring together numerous components into one place. I hope that it will be more than a mere compendium of recent advances in paleontology, genetics, development, ecology, and physiology and how they separately relate to evolution, but rather an integrated synthesis of our understanding of evolution. From the very beginning of life on earth, to the larger and more complex animals and plants that live on its surface today, there has been a remarkable progression, and it is the purpose of this book to seek a deeper understanding of that progression. Why has there been an increase in size and an increase in complexity over time, and especially what is the relation of the genes, the environment, development, and behavior to the extraordinary innovations that have occurred over the last three or so billions of years? Let us begin with a discussion of our current views of the mechanism of evolution.

It is often said that there are two aspects to Darwinism: one is the fact of evolution and the other is the theory of natural selection. It was Darwin's point, and the point of most biologists since Darwin, that the course of evolution can be explained by natural selection; it provides a straightforward mechanism whereby the evolution of plants and animals on the face of the earth can be understood.

In modern terms, natural selection operates on genetic variation; that is to say, those individuals with certain favorable genes and gene combinations will not only survive, but will be relatively more successful in producing offspring, and the result will be that the genes they possess will survive by being passed on to descendants. Even though the selection acts on individuals (or what are known as phenotypes), the ultimate object of the selection lies within the organism in the form of its genes, its genetic constitution (which is known as the genotype). These ideas follow from the work of Mendel, who, in 1866, was the first to obtain a clear insight into the basis of variation with experiments on the crossing of peas. The relevance of this idea to evolution, however, did not come into being with full force until the 1930s with the work of R. A. Fisher, Sewell Wright, and J.B.S. Haldane, who were primarily responsible for the rise of population genetics and what has come to be known as the "new synthesis." By the use of simple mathematical models it was possible to show how the frequency of genes within a population could change with selection pressures of different degrees of magnitude, given the number of generations over which the selection operated. It was also possible to show that the size of the population is an important element: if a population is very small, gene frequency changes within that population can even occur by chance, an idea first clearly stated by Sewell Wright, and called "genetic drift."

In their most recent incarnation, these ideas have given rise to the work of W. D. Hamilton, R. L. Trivers, G. C. Williams, R. Dawkins, and others who see the genes as paramount in natural selection, which operates in such a way that genes appear to be "selfish"; that is, the successful genes are the ones that perpetuate themselves. In Dawkins' words, the genes are the "replicators"; they copy themselves by DNA templates, a process we now understand because of the meteoric rise of molecular biology. He calls the body of an animal or plant (that is, the phenotype) a "vehicle." It is made by the genes through development, and it makes sure that the genes it carries are replicated. In current sociobiology theory these vehicles can go to remarkable extremes to ensure the survival of their genes, such as protecting close relatives that possess many of the same genes. This so-called "altruism" exhibited by the vehicles illustrates the power of gene selection. It is entirely through the mechanistic process of natural selection on genes that organisms can evolve vehicles which protect the genes not only within themselves, but the same genes within the extended family. To some it seems surprising that simple Darwinian selection can result in such complex mechanisms, and there are those who resist these ideas, but this brings us into the realm of the psychology of biologists, which, although a fascinating subject, is not an issue here.

The history of our understanding of natural selection which I have outlined so far has always emphasized the process of selection itself, and the effects of this process are usually illustrated either on a relatively few generations, or on the final result. There has been little attempt to grapple with the great sweep of evolution from the earliest primordial bacteria to the appearance of *Homo sapiens* in recent times. We have been satisfied to say that these principles apply everywhere at each step in the process over the last four billion years or so of earth history. The paleontologist, on the other hand, has been concerned with the major trends of evolution as they are revealed in the fossil record and asks what are the causes of the trends, the shifts, the bursts of speciation, the long periods where no change takes place. Even though all paleontologists today firmly believe in the essential role of natural selection, some have grave reservations about the lessons from the population geneticists, which I have just outlined, because the latter see only gradual change, yet in the fossil record apparently abrupt changes are evident. One aspect of this complex problem is that the population geneticists have been concerned with what happens over thousands of years (at the most), while the paleontologists are concerned primarily with changes that take millions of years; there is little wonder that the rapid change for the latter may seem like a slow one to the former.

What the paleontologists, who have raised this hue and cry, have done for us is to remind us that the genetic mechanism of natural selection is not the only problem; we want also to understand the grand course of evolution that selection is supposed to have produced. It is not enough to understand the principle of internal combustion if one wants to know why there are so many different kinds of motors in the world, capable of so many different functions. Therefore, one must explain not only the rate at which change has taken place over geological time, but also the enormous variety of animals and plants on the surface of the earth today.

There is an interesting blind spot among biologists. While we readily admit that the first organisms were bacteria-like and that the most complex organism of all is our own kind, it is considered bad form to take this as any kind of progression. In the first place, to put ourselves at the pinnacle seems to show the kind of egocentricity that has been a plague to science and fostered such ideas as the earth is the center of the universe or that man was specially created. There is a subconscious desire among us to be democratic even about our position in the great scale of being. From Aristotle onwards there was an idea that there really was a progression towards perfection from plants and lowly worms to human beings. The basis of this scale of being was not evolutionary, but reflected the degree of difference from ourselves, and therefore is unacceptable to us today. In an early notebook Darwin cautioned that we should never use the terms "lower" and "higher" (although he was sensible enough not to follow his own advice), and I have been reprimanded in the past for doing just this. It is quite permissible for the paleontologist to refer to strata as upper and lower, for they are literally above and below each other, and even geological time periods have these adjectives—for example, upper or lower Silurian to mean more recent and more ancient, respectively. But these fossil organisms in the lower strata will, in general, be more primitive in structure as well as belong to a fauna and flora of earlier times, so in this sense "lower" and "higher" are quite acceptable terms. I was raised as a mycologist, a student of fungi and slime molds, and it was the norm to refer to them as lower plants, while angiosperms and gymnosperms are higher plants. But one is flirting with sin if one says a worm is a lower animal and a vertebrate is a higher animal, even though their fossil origins will be found in lower and higher strata. Perhaps plants are forever free of problems of undesirable egocentricity dogma, while all animals are too close to man. I do not know the answer to this subtle question, but in these pages I will treat all animals as I have always treated all plants; firmly, and without favor.

Time

If we concentrate on this grand progression of evolution, let us consider for a moment the element of time. If one looks at the fossil record at different moments in earth history, one finds different kinds of fossils characteristic of different groups of animals and plants. But one must remember that most, but not all, members of an early group became extinct, although frequent extinctions are a significant feature of the fossil past. For instance, the first organisms observed are bacteria-like and are estimated to be about 3.4 billion years old. Bacteria are commonplace today, and therefore we assume that once invented they did not become extinct and that their descendants spanned all those millions of years to the present. From a later time there might be strata in which one finds some invertebrate fossils, but no vertebrate ones, or some algae but no vascular plants. In this way it is possible to give a time sequence to the appearance of the different groups. This has been one of the main concerns of paleontologists, and they have been continually refining their evolutionary time sequences as better information appears on the dating of rocks and as new fossils are discovered.

Unfortunately, the least precise information that is available is for all the earliest lower organisms. Not only are these rocks more ancient, but the fossil record is selective; it preserves best those organisms that have shells or skeletons, and the softer ones often disappear without a trace. It is true that in Precambrian rocks there have been a number of recent important finds of soft-bodied invertebrates, but they are insufficient to build any kind of evolutionary (phylogenetic) tree. Therefore, for both lower animals and plants the relationships between groups and the major modes of construction are largely based on what we know of the structure of relatives of these organisms living today. Fortunately, we do not have to rely entirely on structure but can compare the composition of the organisms' proteins and nucleic acids to measure the degree to which they differ. There is good evidence that changes in key macromolecules occur at a constant rate over time and one can devise a ''molecular clock'' which shows not only how different two living organisms might be, but roughly when they began to diverge in early earth history. It has even been possible, by these molecular methods, to show that there are two fundamentally different kinds of bacteria living today whose ancestors must have diverged at the earliest period of life on earth. Nevertheless, with all these molecular methods, we remain relatively ignorant of the details of primitive evolutionary relationships, and it is still useful to make our phylogenetic trees sufficiently general and vague so that we are committed only to the major trends.

A modern example of such a tree can be seen in Figure 1. The lowest level, which includes bacteria and blue-green algae (or cyanobacteria), is well established. Morphologically these so-called prokaryotes have no nuclei or nuclear membrane, and their DNA is free in the cytoplasm. In this respect, and in other cytological details, they differ significantly from all higher organisms, or eukaryotes. Furthermore, as already pointed out, the earliest fossil organisms known are bacteria (3.4 billion

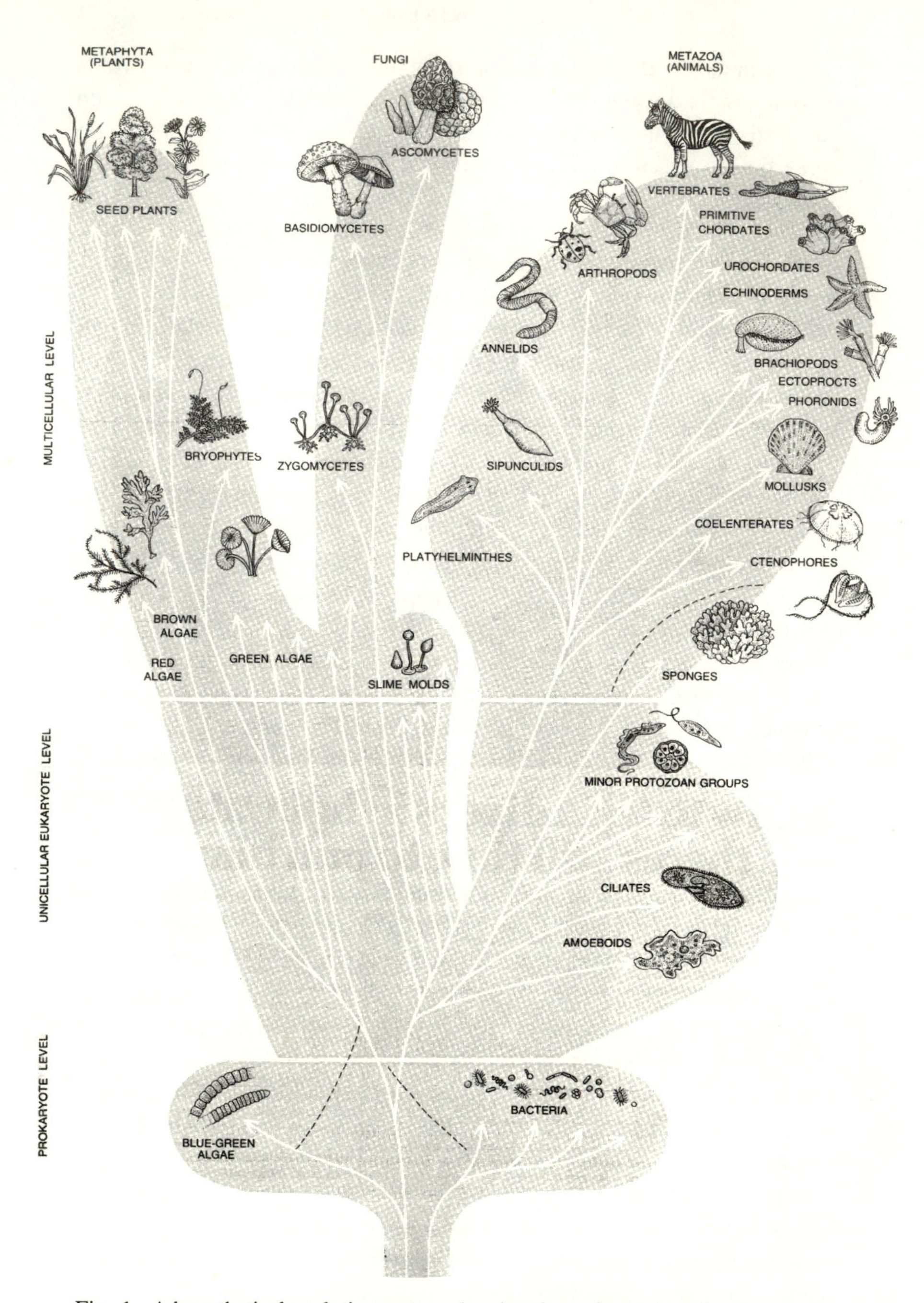

Fig. 1. A hypothetical evolutionary tree showing the early origins of various major groups of animals and plants. (From Valentine 1978, Copyright © by *Scientific American*.)

years ago) and the first known blue-green algae are almost as old (2.7 billion years ago). The time of the origin of eukaryotes is not known, but it is presumed to be more recent. The earliest shelled protozoa (the foraminfera), which could hardly have been the first eukaryotes, are found in rocks over 500 million years old.

During this period, before the invention of multicellularity, there must have been a variety of types of cell construction. The main ones are those of flagellated cells, amoebae, and cells with stiff walls (Fig. 1). We presume the first two gave rise to animal cells since they resemble such cells in their construction, and cells with stiff secondary walls are characteristic of plants. But these three cell types are often mixed, and one finds, for instance, flagellated cells in plants.

If we pursue the plant line, we see a major division between the photosynthetic plants and the nonphotosynthetic plants, the fungi. Again the early fossil record is inadequate, or almost nonexistent, and we must reconstruct from what we know of modern forms. Fungi may have arisen more than once from separate lines, and in morphology they span from simple filaments, as in molds, to great compound masses of filaments, as in mushrooms. There is some evidence to suggest that this span is also an evolutionary progression in time, but it is very weak and requires an unsatisfactory mixture of common sense and faith.

The photosynthetic line is today represented by all sorts of eukaryotic algae (green, brown, golden, and red), which also, no doubt, are not a homogeneous group but had more than one independent origin. Some multicellular green algae, and especially some marine brown algae, reach impressive size, as in the giant kelp, but the much greater size and specialization occurred in the land forms. This specialization is associated with vascular tissue involved in both the transport of substances and the production of fibers to give support. Since the first vascular plants appeared in the Silurian (over 400 million years ago), there has been an excellent fossil record and it is possible to follow all the major groups through time (Fig. 2).

If one looks to the phylogenetic tree of animals, one sees a similar pattern. It is difficult to know the sequence of appearance in time of the various invertebrate groups, largely because of the bias towards the fossils with skeletons. On the other hand, vertebrates have left an excellent record—one that can be reconstructed in great detail. From it one can see that fish appeared first, amphibians considerably later; these were followed by reptiles, and later mammals (Fig. 3). At a somewhat later

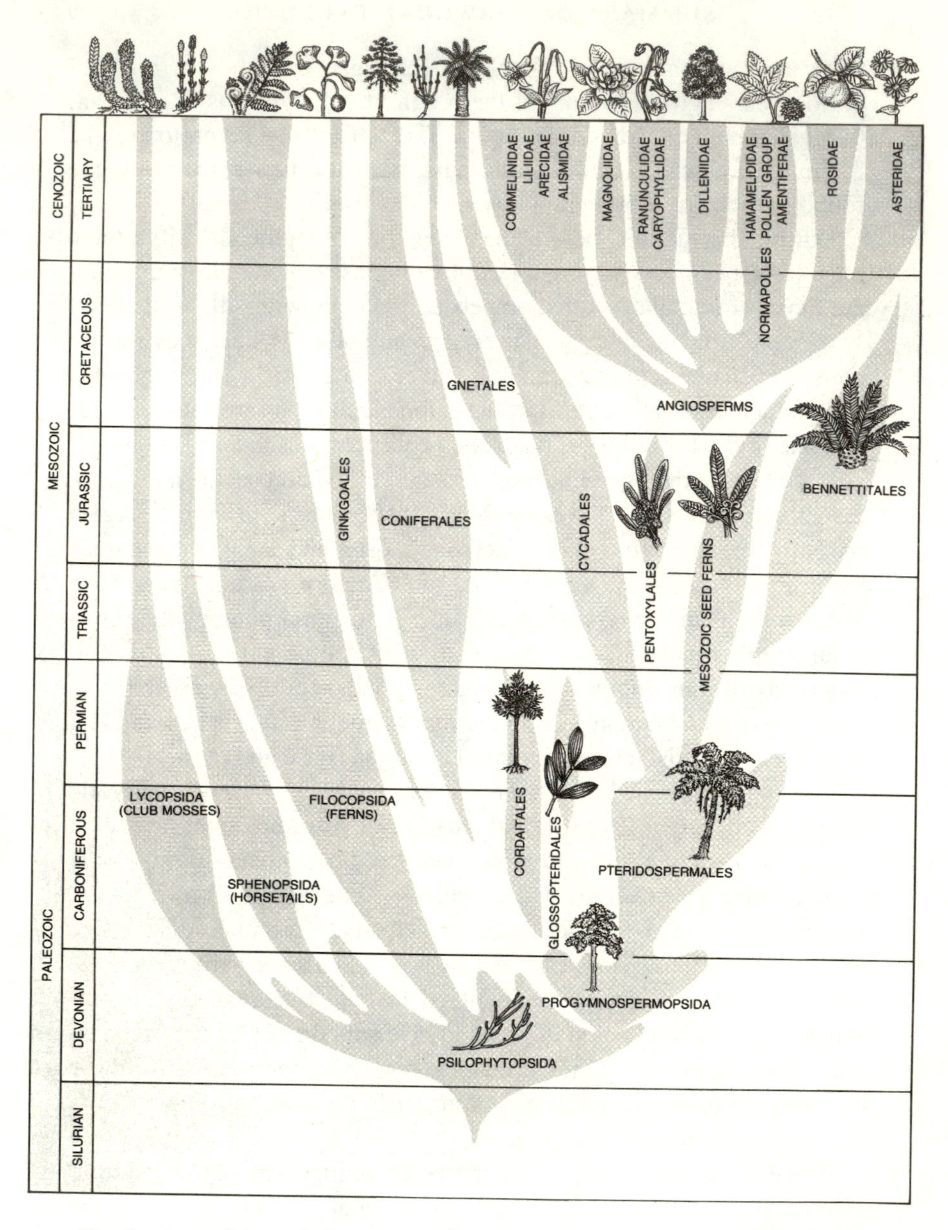

Fig. 2. A possible evolutionary tree of the evolution of plants from the Devonian to today. (From Valentine 1978, Copyright © by *Scientific American*.)

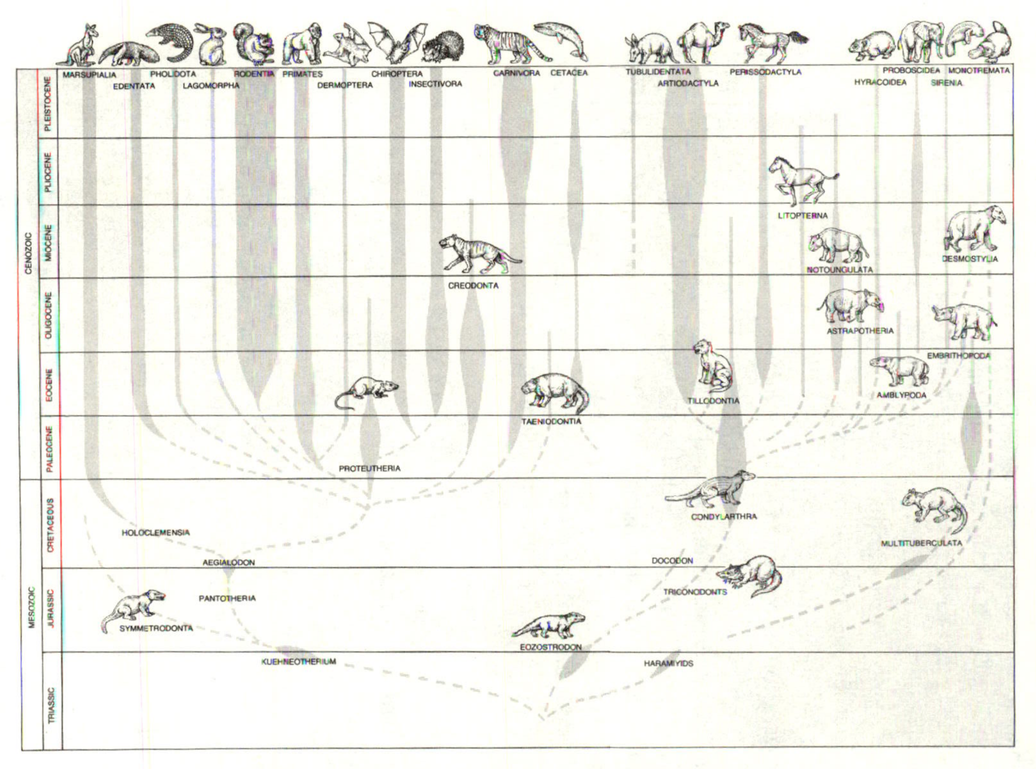

Fig. 3. An evolutionary tree showing the history of mammals. The early forms were all small and coexisted with dinosaurs, including some of the first primates. (From Valentine 1978, Copyright © by *Scientific American*.)

date, after the appearance of the first mammals, birds evolved from reptiles. Note that the first primates appeared quite late, in the same era when we find the extinctions of the dinosaurs.

Later in this book we will have occasion to stop at many places and examine specific steps in this large picture. Here I want to paint the evolutionary sequence with a broad brush. Since we know so much more about recent fossil history and find very early periods less clear, it may be useful to put the time of formation of the major groups of animals and plants on a logarithmic scale where recent epochs are magnified and earlier ones become progressively compressed (Fig. 4). A mere glance at this figure makes the point that I wish to emphasize here: there is a progression in evolution of both plants and animals. To anticipate what is to come, we will examine in detail the nature of this progression. Is it a progression in size and a progression in the complexity of organisms, and if this is so, what is the reason for it and what are the consequences for the organisms at the different levels? Before addressing ourselves directly to the latter questions, it is important to consider carefully the nature of the progression; why has there been a progression in the first place, and what are the elements that shape it?

Natural selection

The standard and correct answer that one would receive from any biologist today is that the sweep of evolution is the result of natural selection. This is certainly the prime mover in all of evolution. It may be that there are important additional factors that play a role, but no one questions the great significance of natural selection. Because it is so important, let us discuss it in more detail so that it is clearly understood what we mean by it.

The point has already been made that natural selection acts ultimately on genes. This simple fact cannot be overemphasized. It is easy to show that selection cannot act on the level of morphology alone. If through human intervention or some natural cause, one body, one phenotype is altered, then the effect will not be passed on to the next generation. For instance, one can dock the tails of all the sheep in a flock, but this radical selection only lasts one generation; nothing is passed on to the offspring. The intermediate case is less clear-cut and has led to considerable confusion. Animals are capable of behavior, and some behavior patterns could be under selection pressure. Imagine, for instance, that

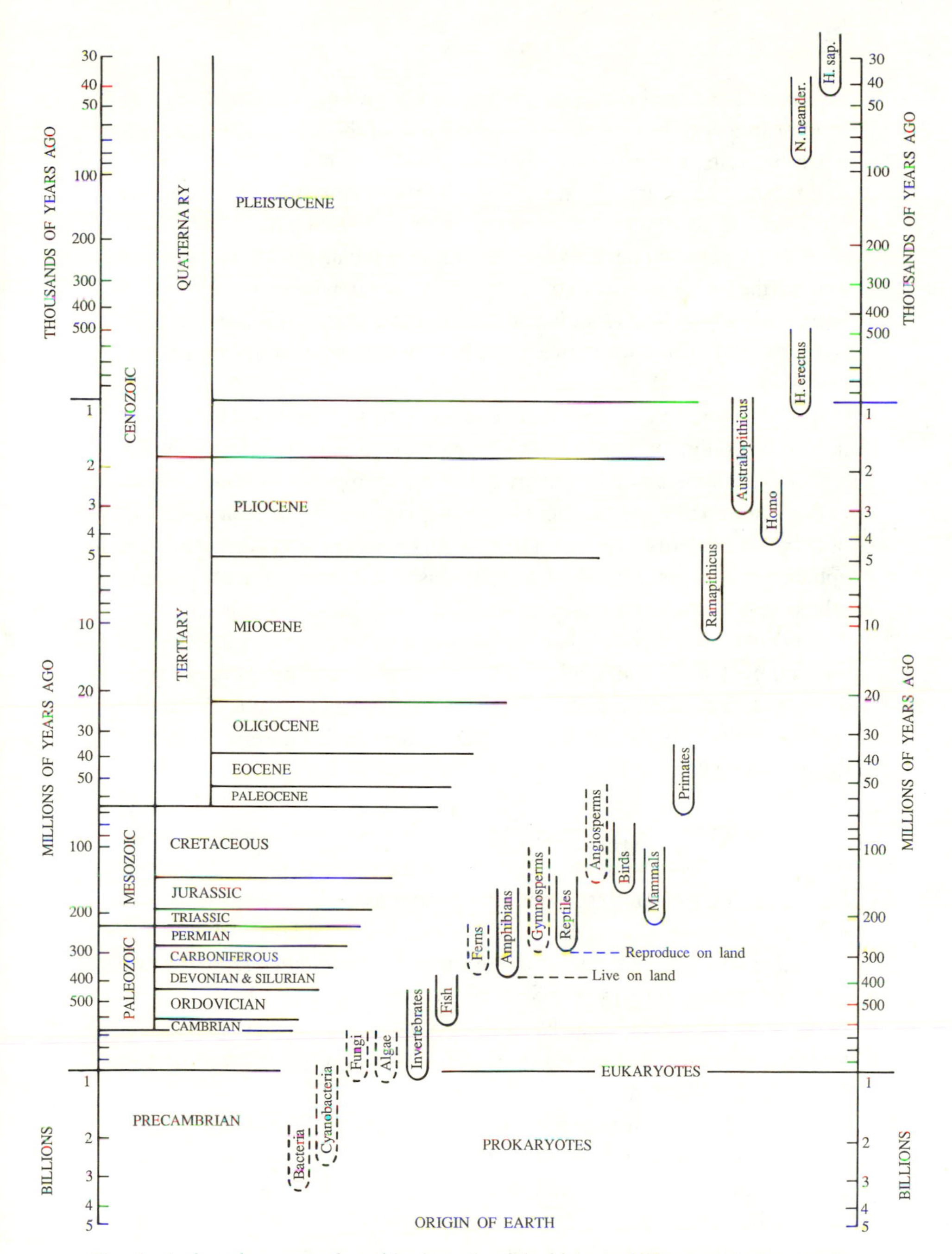

Fig. 4. A chart drawn on a logarithmic scale of the history of life on earth showing the time of the beginning of each major group of organisms, from bacteria to man. Plants are indicated by dashed lines; animals by solid lines. Note that the logarithmic scale greatly compresses the changes that occurred long ago, as can be seen from the scales on each side of the chart.

the use of some tool is helpful for feeding in chimpanzees, either in the wild or in a zoo. Let us assume that the use of the tool is not inherited, not a behavior pattern determined by genes, but one that is passed on by imitation and learning. In this case the tool-using trick will be passed on from one generation to the next, but it will be behavioral transmission, not a genetic one. Despite the fact that it can span generations, it is nevertheless not the kind of selection we mean when we use the term natural selection. The latter is confined to genetic transmission and there is a profound difference between the two. It is important to explain why this is so.

The appearance of a new gene and its transmission to other individuals in a population, as we have learned from the population geneticist, is a slow process and takes many generations in order to become fixed in a population, even in a small population where such changes can occur more rapidly. Behavioral transmission, on the other hand, is ephemeral and can come and go with ease. A new behavioral pattern, which may be highly advantageous to the individual in a particular environment or set of circumstances, may last as long as needed; but if the environment should suddenly shift, the pattern could be dropped or altered within a generation or in a much shorter span of time. Were the pattern genetically determined, a change needed because of a shift in the environmental conditions could take many years and many generations.

Because Darwin did not understand the basis of heredity, since the work of Mendel was not appreciated until the beginning of this century, he was primarily concerned with the problems of Lamarckism. This is the idea that somehow a change in environment could directly effect an organism so that its genetic constitution becomes altered. Despite enormous efforts on the part of many people there has never been any evidence that this can happen. Even if a convincing case were discovered tomorrow, it would still be true that this is not a general mechanism involved in evolution.

If natural selection is confined to gene inheritance, we can see that indeed gene perpetuation is the prime result of this selection, never forgetting that the genes are contained in the "vehicle," for it is this phenotype they create to protect themselves that helps them pass from one generation to the next. The whole process is entirely mechanistic: a particular environment will select for particular phenotypes and this will result in a change in the frequency of some genes which in turn will affect the future phenotypes of a population. In this sense, it is impossible to separate genotype from phenotype. The phenotype is first affected by selection, and the new gene arrangements direct the appearance of the subsequent phenotypes, which are built anew each generation.

There are three enormously important consequences of this interplay between the phenotype and the genotype during the course of evolution caused by natural selection. For the moment I am putting to one side the question of why there has been a progression from small to large, from simple to complex, and am concentrating on the consequences. It is true that one cannot be discussed without the other; it is merely a matter of choosing a convenient sequence. One of the three consequences of our evolutionary progression is that organisms have increasingly elaborate development or embryology as their phenotypes become more complex. The second is that as the evolutionary progression occurs, the environment becomes populated with a greater variety of animals and plants, with larger ranges of sizes and complexities, and this alters the environment so that the directions of the forces of natural selection in turn become modified. Finally, in animal evolution, there has emerged a new property, which we call behavior. It involves teaching and learning in its more advanced form, and this also has had an exceedingly important effect on the evolution of higher vertebrates. Let us now examine briefly these three consequences of evolutionary progression, which we can label development, ecology, and behavior.

Development

The idea that the development of an organism relates to the evolution of its ancestors is an old one. It goes back at least to Karl Ernst von Baer fairly early in the last century; it played a significant role in Darwin's thinking, and was brought to a peak in the nineteenth century by that great enthusiast, Ernst Haeckel. The ideas centered almost exclusively around the evolution of vertebrates, and it was clear that early embryos of fish and more advanced vertebrates, including man, had very similar early stages of development, but differed progressively as they matured. This was the insight of von Baer, and Haeckel produced the general principle that ontogeny (development) recapitulates phylogeny (evolution). While there is much truth to his "biogenic law," it is sufficiently oversimplified to have caused considerable irritation in the twentieth century, with the result of generating some good ideas about the relation of development to evolution. The modern pioneer was W. Garstang (1922), who was the first to suggest that since embryonic stages, and especially larval ones, may themselves be well adapted to their environment, selection can operate at different stages during the course of the life history of an organism. G. R. de Beer (1958) took up the anti-Haeckel torch and added to Garstang's list of instances where recapitulation was only one way in which descendants could be modified from their ancestors; there were many other possible shifts in the timing of development or heterochrony (a term originally coined by Haeckel).

It is an interesting historical fact that in this century the great advances in evolutionary biology made possible by the population geneticists completely ignored the role of development. The reasons for this are largely because embryologists and evolutionary biologists were concerned with totally different questions. Therefore the mainstream of evolutionary studies was concerned only with the evolution of adults, and development seemed to have little bearing on the central problem.

There were a few biologists who were exceptions. For instance, C. H. Waddington (1940, 1957) fully appreciated the fact that it was the development that was responsible for the adult and that genetic change, in order for evolution to occur, had to modify the development. In a different way, I. I. Schmalhausen (1949) put a similar emphasis on the role of development in molding the phenotype, and consequently the significance of development in evolution. Furthermore, both were interested in the stability of developmental pathways, and Waddington devised a concept of canalization where the gene-controlled development seems to be buffered and adjusted so that it would proceed in one way only, despite environmental disturbance. In later studies Waddington (1957) became interested in ways in which the environment could directly affect gene expression; and he was able to show in experimental studies of the fruit fly, *Drosophila*, that if he selected for a phenotype that appeared under certain environmental conditions (a pattern of the veins of the wing that would appear when the developing fly was given a heat shock), after a number of generations of selection of the most sensitive flies, he would be able to produce cross-veinless flies without the heat shocks.

Even though the ideas of Waddington and, to a lesser degree, those of Schmalhausen were often referred to, they nevertheless did not fit into the mainstream of thinking on the subject of evolution. This tension between development and evolution seemed to stem from the fact that there was no place where they could comfortably meld to make a comprehensive synthesis, a situation that is slowly beginning to change. It is this change that is one of my prime reasons for writing this book.

In the meantime, biologists for the last fifty years have been reading with interest and profit the book (and its later editions) of G. R. de Beer (1930 et seq.) on the various manifestations of heterochrony. His classical education stimulated him to indulge in the most formidable array of new terms to describe the various possible kinds of heterochrony, but despite these obstacles his important message emerges. It is possible to advance or retard some developmental stages, or even change the sequence of developmental events, and these shifts in developmental timing will have significant consequences for the morphology of the emerging adult. Later we will examine heterochrony in detail; here I want merely to make the point that major differences in the structure of descendants are possible as a result of minor shifts in the timing of development. In this way it is possible to produce quite radical changes in organisms that involve relatively few gene changes. I anticipate here what is to come only by saying that the main reason the ideas of Haeckel, Garstang, de Beer, and others have become so popular is that in recent years many biologists have begun to ask how major (versus minor) changes in evolution can take place, and one answer is to be found in heterochrony.

The form that the new interest in the relation of development to evolution has assumed is still indistinct. It has been generated by a number of new trends, and let me mention them briefly.

The first comes from molecular genetics. The progress in that field, ever since the rocket launched by Watson and Crick in 1953, has been nothing short of astounding, and seems to continue without any loss of momentum. The most recent discoveries have a direct bearing on evolution in that the evidence now points to the possibility that the classical ideas of genetic change coming through point mutation and recombination have become greatly enlarged. Not only is it possible for major rearrangements to occur within a genome—the so-called gene jumping—but also it is possible for small bits of genetic material to shift from one individual to another (transformation). This clearly broadens the possibility of significant genetic change taking place rather abruptly.

A second major cause of a resurgence of interest in development in evolutionary studies comes from paleontology. S. J. Gould and N. Eldredge (1977) stirred enormous interest (and an equal amount of controversy) in suggesting, from the fossil record, that one was forced to the conclusion that evolution occurred in fits and starts: sometimes there were long periods of no change in fossil lines, and at other times there were sudden bursts of new species. This evolution by "punctuated equilibria," as they called it, produced a turmoil in that it seems to be pitted against the gradualism of population genetics. That this was not so became drowned out in the heat of the argument: disagreements are more fun than agreements. But one of the beneficial effects of the punctuated equilibrium position was that people again began to worry about major changes occurring rapidly during evolution. The need for explanations of so-called macroevolution was again in the fore. I say "again" because this had been a major preoccupation of Richard Goldschmidt (1940), who coined the catchy phrase "hopeful monsters." He envisioned that there occurred during evolution "megamutations" and that these were responsible for the appearance of new major groups. These ideas of Goldschmidt were not very popular among biologists at the time, and as my colleague Henry Horn has pointed out to me, they already had been elegantly refuted by R. A. Fisher (1930:41–45). They have, however, resurfaced in a more sophisticated form.

As a result the paleontologists and the molecular geneticists have become unexpected allies in a new search for mechanisms of rapid evolutionary change. While they are operating at levels that could not be further apart, the connection between the two was made forcefully by a paper of M.-C. King and A. C. Wilson (1975). They showed that on the basis of differences in certain proteins the genetic similarity between chimpanzees and human beings is remarkably close, no greater than what is found between closely related species of *Drosophila*. Yet we feel that chimpanzees are enormously different from ourselves. Therefore the conclusion from this landmark paper is that some gene changes must have a much more significant effect on the phenotype than others.

This immediately raises the question of what kinds of gene changes produce such profound morphological changes. There are two answers. One is that they are probably regulatory genes, that is, genes which affect a whole battery of other genes and in this way produce a large effect. (This is an idea first advanced by R. J. Britten and E. H. Davidson in 1969.) The other is that those genes could affect the timing of development. It has been well recognized since the pioneer work of L. Bolk (see Gould 1977) that human beings are neotenous apes, that is, our embryonic development has been greatly lengthened so that as adults we still retain many features found only in the fetus in apes. Therefore, if a few genes were responsible for affecting the timing of development, one could expect major changes in the adults during the course of evolution.

Ecology

Another approach to evolution that has been actively pursued also had its origin in the last century. This is the study of ecology, the natural setting of animals and plants. Both Darwin and Wallace had a profound understanding of it, for not only were they gifted naturalists, but they realized that natural selection always operated within an environment; it was not possible to separate the process of evolution from the surroundings in which it occurred. It is remarkable that when we read their works today, they seem to have such a modern insight into how selection pressures might be operating in a natural habitat.

In this century the first wave of new concern came from C. S. Elton (1927) and others who were interested in the structure of communities and began our understanding of food webs and food pyramids. This started a movement towards a more quantitative analysis of the environmental forces, which was greatly advanced by G. E. Hutchinson and came into full bloom under the enormous influence of his student Robert MacArthur. MacArthur was initially trained as a mathematician, and he sought simple models which would give deep insights into the way animals and plants organized themselves in a community. But, like Darwin and Wallace, he was also a gifted naturalist and constantly cautioned that no model should be dogmatically accepted, but should always be considered something capable of improvement with the accumulation of further knowledge. The revolution he produced was to leap from the general attitude in the mid-fifties that the striking thing about communities was their complexity, to the realization that although this was understandably true, there were certain straightforward mathematical models that showed beautiful, simplified relations lying hidden within the complexities. Indeed, one of the patterns he sought to understand was the differing degrees of complexity in different communities. He favored calling this new approach "population biology" to distinguish it from the more traditional ecology, and as one might expect it was met with enormous resistance: it was oversimplifying the complexity of nature in the unfriendly clothing of mathematics. But the success of population biology has been so remarkable in the last twenty years that the revolution is over.

Although many of these points will come up later, let me say here a few general things about the relation of ecology to evolution. In the first place, the environment of any one plant or animal is made up in part by the physical elements and in part by the other animals and plants that surround it. Therefore the relation between organisms is one of the important factors which affect the direction of selection. This may seem tremendously obvious, but nevertheless it is sufficiently important to stress here.

A consequence of any one organism being surrounded by others is that as natural selection changes those surrounding organisms through modification, the environment is correspondingly altered. Therefore, one of the results of natural selection is continual change in the environment, and that means probable changes in the direction of the selection. In this sense natural selection in an ecological setting is unstable because continual change through selection is inevitable. Equilibrium could only be reached when all the forces of selection move to keep the status quo. But such stasis for a whole community, while theoretically possible, is hard to imagine. It is more probable that one adaptable species is stable under changing environmental conditions for long periods of time. Putting the whole ecosystem on hold probably never occurs, for even the climate will tend to push and shove in unpredictable ways.

Since in any one habitat there is a limited supply of energy, be it sunlight or organic food, inevitably there is enormous competition among organisms for those resources. In recent years, following G. C. Williams (1966), we have come to appreciate the fact that competition is between individual organisms and not between species. It may be that effective or inadequate success of individuals in gaining food leads to the threat of extinction of a species, but species are only indirectly competing entities.

Again Darwin was fully aware of the key role played by competition in driving natural selection. The successful individual animals that exploited food were the ones that reproduced; and the plant that caught most of the sun for photosynthesis prospered, while those forced into the shade of a larger plant waned and produced few or no seeds. One of the prime moving forces behind natural selection is competition. If the efficiency in getting a larger share of the food is increased, so will productivity, and those genes that were responsible for the increased efficiency will remain fixed in the population. Other competitors will also have genetic variation subject to selection, so that they can either hold their own or perhaps even outcompete the original dominant competitor. This situation will turn into what R. Dawkins and J. R. Krebs (1979) refer to as an "arms race," and clearly will play a major role in guiding selection.

There is another especially potent form of competition, which is predation. It does not involve the competition of individuals of one species, between rough equals, but between members of different species, between the oppressors and the oppressed. Predation is a major driving force in evolution, where the predator is selected for effective attack and the prey is selected for effective escape. Furthermore, when the level of predation is high, the prey often become sufficiently decimated so that competition among themselves vanishes; their survival depends entirely on their relation to the more demanding predators.

Whether the struggle is between species or within a species, there is an escape: an animal or a plant can pull away to a new environment and in that way avoid predation and competition. Perhaps it can find a location where its competitor or oppressor is absent, either by chance, or because for some reason it cannot penetrate the area. One could give many examples: it invades uninhabited islands or forbidding climates; or it becomes so large that it becomes an effective competitor or a prey too large to be subdued; or it becomes so small that it can effectively hide from a predator or make use of a food which is unavailable to, or difficult to eat by, the larger competitor. In the case of plants, some smaller shrubs have perfected a way of growing successfully in the shade; they will flourish in the understory of a forest and manage with a lesser amount of light. Therefore, one result of predation and competition is to drive organisms into new worlds to avoid harassment, and this "pioneering" effect will result in a great increase in the number of kinds of animals or plants that exist in any one geographic region, a phenomenon known as diversity, or species diversity. Pioneering is responsible for the invasion of especially inimicable habitats, such as polar regions or the great depths of the sea; it is an important force in the distribution and diversity of plants over the globe, and a correspondingly important force in evolution.

There is a third effect of competition, besides the arms race and pioneering, that comes under the general category of "if you can't lick 'em, join 'em.'' There are many beautiful examples in which organisms live together in some kind of arrangement that benefits at least one of them. It is true that a great many of these instances of parasitism or mutualism do not involve the coming together of organisms that were in direct competition, but rather quite divergent organisms that gain by the association. For instance, the cellulose digesting microorganisms that inhabit the guts of termites or cows are certainly partners that were never in competition. An example of competition turning into an association is more subtle and usually benefits one of the partners and not the other. Scavengers fill this category; they may pacify the prime predator so that they become true parasites. Good examples are found in ant societies, where other parasitic species of insects can, by producing the right odor or making the right gestures, fool the ant into thinking they are fellow ants and are dutifully fed by their hosts. There is the possibility that many cases of parasitism known today arose when one competitor moved in on the other. Note that this invasion constitutes a rather specialized form of pioneering, and like the instances discussed previously, is the direct result of competition.

Behavior

The point has already been made that there is a fundamental distinction between genetic transmission of information and behavioral transmission. Furthermore, there is an important connection between genes and behavior; they can be related in two ways. Some behavior patterns are genetically fixed and handed down in the most rigid form. One needs only be reminded of the offspring of a solitary wasp that emerges as an adult from its nest, never having seen another wasp, yet it can forage for food, mate, and perform all its living functions without a lesson. Obviously all these behaviors have been directly inherited; they are ''hard wired,'' to use the jargon of the neurobiologist. But other animals, such as ourselves, need prolonged teaching and learning in order to do many of the things we do as adults in our daily lives. In these cases it is the capacity to teach and learn that is inherited, and the actual information is passed by behavioral transmission. Between these two extremes lies a continuum of cases where varying proportions of hard wiring and flexible learning occur. The variability of the proportion of fixed behavior to flexible, learned behavior is enormous even in one group of organisms. For instance, in some birds the entire song is built in, so that singing perfectly requires no hearing of other individuals, while in other species the song is learned to varying degrees from older birds. The fact that birds are so variable in this respect suggests that the two forms of transmission are adaptive for different reasons, and the differences found among similar organisms is to a large extent the result of natural selection, a matter that will be considered in detail in Chapter 7.

In the evolution of animals, behavior has obviously played a significant role, and that is the relevant point here. By developing elaborate behavior patterns, animals can compete more successfully; they have a greater variety of ways in which to cope with external exigencies. Although each case is special, the greater the flexibility of response, the more effective can be the food catching or the predator avoiding. Therefore, increased reliance on behavior is one very successful way to arm in the "arms race."

The rise of elaborate behavior patterns has played another role in evolution. Earlier the point was made that organisms are changed through evolutionary time, and this in turn changes the environment and alters the nature of the direction of natural selection in large and in small, subtle ways. With the appearance of behavior, these changes in the environment which link to natural selection became more influential. A slight shift in behavior can produce all sorts of ripple effects in a community that will have important results affecting the selection pressures of other organisms. A new method of hunting by a major predator will, for instance, produce new pressures on the prey, and some of these defense mechanisms may be genetically fixed so the change will be by natural selection. The most extreme example will be found in human beings, who through their mental capacities have invented artificial selection for domestic animals and crop plants. This illustrates the ultimate in the power of behavior on evolution.

There is another relation between behavior and development that must also be mentioned here. It is a subject that has fascinated me for a long time, and there are many others today who are contributing to it in important ways. Behavior has a development along with the changes in morphology of the animal, and the stages of both have many interesting similarities. For instance, in both there are sensitive periods during which the environment can affect the direction of the future development. It is true that one comprises the development of the whole body, while the other involves the development of the nervous system specifically, and for this reason it is not surprising that there should be many similarities between the two. The surprise comes, rather, from the fact that the end result is so different. In one case, the result is a functioning structure while in the other it is a pattern of behavior, and these not only differ in substance, but in their consequences to evolution.

Genetics, development, ecology, behavior, and evolution

The fundamental message I wish to impart in this introductory chapter is that evolution can no longer be looked at solely as changes in gene frequencies within populations, or as fossil lineages: it is now essential to consider simultaneously the roles of genetics, development, ecology, and behavior. Only then can we begin to see the full dimensions of evolution by natural selection. This is merely repeating precisely the position of Darwin, for in his *On the Origin of Species* he used variation (genetics), embryology (development), geographical distribution (ecology), and instinct (behavior), along with the fossil record, as sources of evidence for evolution by natural selection. What has happened in the meantime is that, as biology has become so highly specialized, we have tended to emphasize one or the other of the components and not look at them all together. Recently there has been a movement towards some grouping, and there have been discussions of evolution and ecology, or evolution and behavior, or evolution and development. As we have just seen from our superficial glance, these different disciplines have shed new light on evolution, but the boundaries between them have been sharp. What I would like to do in this book is to go back to the perspective of Darwin and look at all these aspects of evolution together, for there are connections between genetics, development, ecology, and behavior, and these connections themselves may give us a deeper insight into our understanding of evolution. It is not my intention here simply to rephrase what Darwin said in his great book, but to look at the subject again from a comprehensive view, taking advantage of the enormous strides that have been made in the disciplines of ecology, genetics, development, and behavior.

Nor do I intend to discuss each one of these subjects separately, as I have very briefly in this chapter. Rather, the aim will be to isolate two important generalizations about evolution, and then see how the four disciplines relate to those generalizations. The two generalizations are that during the entire courses of evolution, from the first appearance of bacteria in primeval mud to the fauna and flora of today, there has been (1) an evolution from small to large, and more importantly (2) an evolution from simple to complex. Therefore, our task will be to reexamine this reincarnation, in modern form, of the great chain of being from the point of view of current population biology, genetics, developmental biology, and behavior.

This odyssey through time gives one a deeper insight into the basic understructure of the hierarchical levels that are the organization of life. We will see how the levels arose and how they are related one to another. For instance, the mode of inheritance differs with the level. Also, the degree of complexity within an individual organism has a relation to the complexity between organisms within an ecological community. In addition, we will see that in the evolution of complexity there are two opposing tendencies, both encouraged by natural selection: the tendency for living structures to become integrated into units, and the tendency for those units to become separate and isolated.

I shall not anticipate all that is to come, but I hope that this interdisciplinary view of evolution will bring forth new truths that have not been immediately obvious. The truths of a century ago have not waned; instead, they have taken on depth and glitter in the light of our new knowledge. Yet it is hard to see the interconnections between the contributions of different disciplines, for we have tended to consider each discipline separately. Here we will look at everything together to seek the essence of modern evolution.

Further References

V. Hamburger (1980) provides an excellent discussion of why, in the first half of this century, evolutionary biology, and especially proponents of the new synthesis, ignored embryology. However, embryology has played an important role in evolutionary thinking, starting long before Darwin, as S. J. Gould (1977) shows in detail in his fine book.

Chapter 2

Evidence for the Evolution of Size Increase (and Decrease) from the Fossil Record

ONE OF the more important generalizations about evolution is that there have been great variations in size increase and decrease during the course of time. Since the size of organisms has such significance with respect to their complexity, we will consider size-related subjects first and then, by Chapter 5, enter fully into the problems of biological complexity.

In this chapter I will show that there is an overall trend towards size increase during the course of evolution. By this I mean that there has been an increase, over time, of the upper size limits of both animals and plants. It is easy to describe this trend, as will be done first, but it is much more important to explain it. We shall begin with a possible explanation in this chapter, but really come to grips with the problem in the next.

First, an important note about size in organisms. A distinction can be made between an individual organism that reproduces sexually in discrete generations and one that buds off asexual offspring which may remain attached, such as a coral, or many higher plants that remain connected through their root systems. Such asexual, budding forms are called "clonal" organisms because the vegetative buds are likely to be genetically identical, while a sexual organism with separate generations is called "aclonal." In the immediate discussion that follows I shall be talking about size in aclonal plants and animals, but clonal organisms, by their repeated budding (if the buds remain attached), can be enormous. For instance, huckleberry is a clonal plant, and one patch of connected bushes has been measured at almost 2,000 meters in diameter (and estimated to be 13,000 years old!). I will return to this interesting distinction between organisms with these two different kinds of life histories.

The first known fossils are minute bacteria-like organisms found in rocks that date back as early as 3.4 billion years ago, while today we have the largest animal ever known, the blue whale, and the largest plant, the giant sequoia. Therefore, in the broadest sense, there has been an evolution of size increase during the course of evolution if only the maximum sizes attained at any one geological period are considered (Fig. 5). In over 3 billion years the increase has been more than seven orders of magnitude, from micrometers to a hundred meters of increase in length. This represents a formidable magnification of the size of the gene-bearing vehicle.

The difficulty with this trend is that, although undoubtedly it is real, it is nevertheless a very crude generalization. Because of the gigantic gaps in the fossil record, there are only three points on the graph between the early bacteria-like cells and the well-preserved forms in the Cambrian, a span of 2.8 billion years. One point consists of fossil colonial blue-green algae that resemble bacteria in their cytological structure (i.e., prokaryotes) but form chains of cells, and because they accumulate calcareous deposits they have been preserved. Another point is for the first fossil eukaryotic organisms (called acritarchs). We are hardly

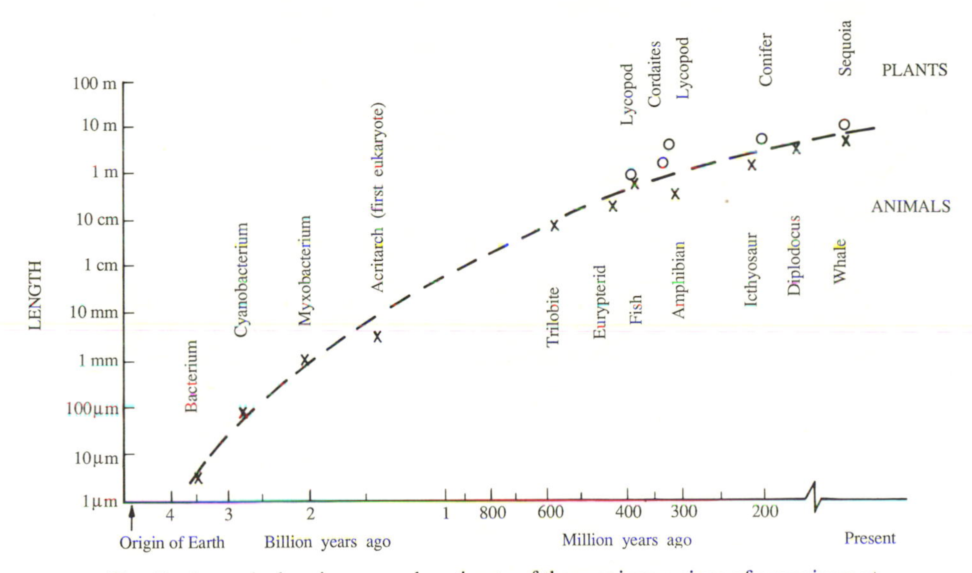

Fig. 5. A graph showing a rough estimate of the maximum sizes of organisms at different periods of life on earth. Note that both the length (or height) of the organisms and the time are on logarithmic scales. (Modified from Bonner 1965.)

in any position to do much more than plot a very questionable dashed line between the first fossil bacteria, colonial blue-green algae, acritarchs, and the invertebrates around the Cambrian period. Even where we do have good fossil records in the last 600 million years, there is still wide variation in the maximum sizes for any one group. Furthermore, length is in some cases only a moderately good index of size. For instance, *Diplodocus* is indeed the longest dinosaur (27 meters), but it is not as massive as *Brachiosaurus*; the estimates for their weights are 35 tons for *Diplodocus* and 85 tons for *Brachiosaurus*. Another difficulty with Figure 5 is that the bulk of vascular plants is dead tissue—fibers and tubes that make up the wood. This is growth by accretion and has no exact counterpart among vertebrates. A parallel among animals might be a snail that builds an ever increasing nonliving shell. However, in both snails and large trees there is also a steady increase in the amount of living tissue during the course of its development. Consequently, length has a different meaning for higher plants than for higher animals, and this difference is reflected in Figure 5.

It must be emphasized again that these are the upper size limits or record sizes for each epoch. Organisms of intermediate size between minute bacteria and the largest animals and plants existed at all times, and certainly do today, a matter we will consider in detail presently.

In the case of clonal organisms it is clear that in the early paleontological history of both animals and plants there was a significant trend in size increase by clonal growth. The earliest forms were predominantly clonal; this was an easy and inexpensive way of increasing size while avoiding the problems of complexity, a matter which will be our major concern in later chapters.

For aclonal animals it was recognized quite early in the study of paleontology that certain fossil series showed a trend of size increase over a period of time. It was E. D. Cope in the last century who saw the point clearly and made a generalization that has become to be known as Cope's rule or law. This rule says that in certain vertebrate lines there is a general trend towards size increase over geological time. For instance, this may be seen in Figure 6 in the evolution of therapod dinosaurs and how they changed over a span of many millions of years. Camels, elephants, and horses provide other well-known examples.

It was not until the middle of this century that N. D. Newell (1949) showed in an elegant paper that the same principles apply to invertebrates. Among foraminferans, arthropods, echinoderms, brachiopods,

Fig. 6. Theropod dinosaurs of a range of body sizes. The bones of the larger animals are relatively thicker. From the smallest to the largest, these animals (all representatives of different species) are a 165-kg ornithomimid, a 735-kg tyrannosaurid, a 2,500-kg tyrannosaurid, and a 6,000-kg tyrannosaurid (*Tyrannosaurus rex*). (From McMahon and Bonner 1983, Copyright © *Scientific American Books*.)

and ammonites, all widely separated groups, such trends of size increase also occur; they obey Cope's rule in a way that seems exactly parallel to what is found among vertebrates. This is illustrated in a striking fashion in Figure 7, where we can see that not only the duration of these trends in size increase is in the same order of magnitude (they span periods from roughly 5 to 50 million years), but the rates of size increase are of the same order of magnitude, as shown by the slopes of the lines in Figure 7.

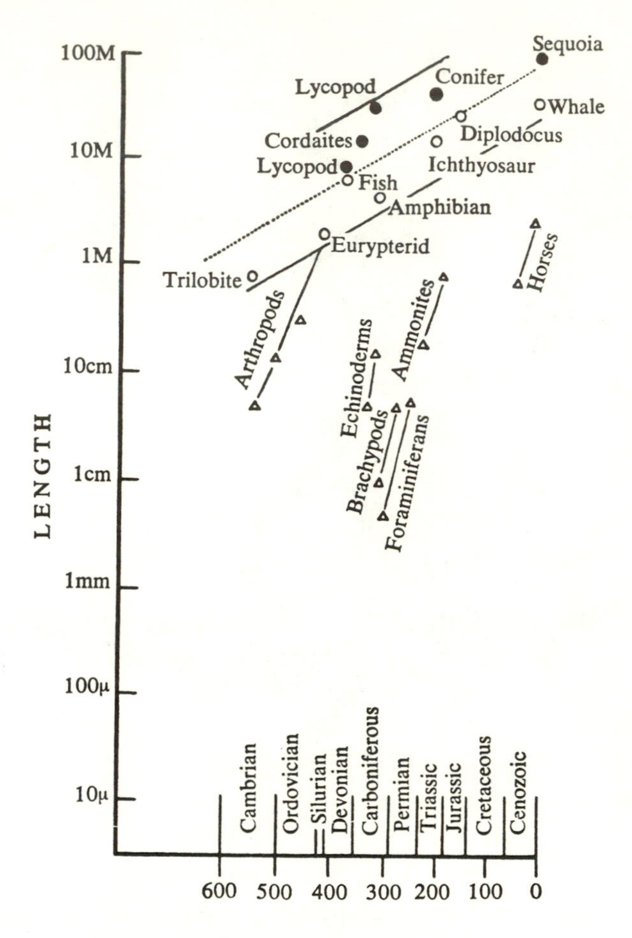

Fig. 7. A graph showing the relatively short-term trends of size increase in various animals. Above these examples one can see some of the maximum sizes of some animals and plants reached in various geological periods (see Fig. 5). (From Bonner 1965; data on invertebrates is from Newell 1949.)

In the last few years we have come to see Cope's rule in a different way. The foundation of this new understanding was laid down by B. Kurtén, P. D. Gingerich, and others who showed that if one looks at fossil sequences one often finds periods of size increase followed by periods of size decrease. In other words, a close, more finely tuned inspection of the fossil record shows that there is as much getting smaller as there is getting larger. Let me give a few examples.

For instance, B. Kurtén (1959, 1968) measured the second lower molar of fossil bears from the Pleistocene (for molar size is known to be a good index of body size). Over a range of 450 thousand years, as can be seen in Figure 8, there are successive periods of increase and decrease in the size of the bears. In another excellent example, P. D. Gingerich (1974) followed the molar size of a condylarth (a primitive hoofed mammal) through a series of stratigraphic layers (amounting roughly to a span of five million years) and again found trends of getting smaller as well as larger (Fig. 9). One final example may be found in a radiolarian, a small, shelled protozoan that lives in the ocean, over a period of approximately a million years (Fig. 10). Note that in all these cases a lineage is being followed in successive strata which give a temporal sequence, and in this way the size meanderings can be carefully traced.

Examples of this sort do not seem to fall in with Cope's rule, and the reason for this is neatly explained in an important paper by Gingerich (1983). He brings together a great mass of data gathered from many authors on size changes over different time spans: from selection experiments that require less than ten years, to paleontological sequences that span hundreds of millions of years. While there is a considerable scatter in the data, if he plots the rate of size change against the time over which

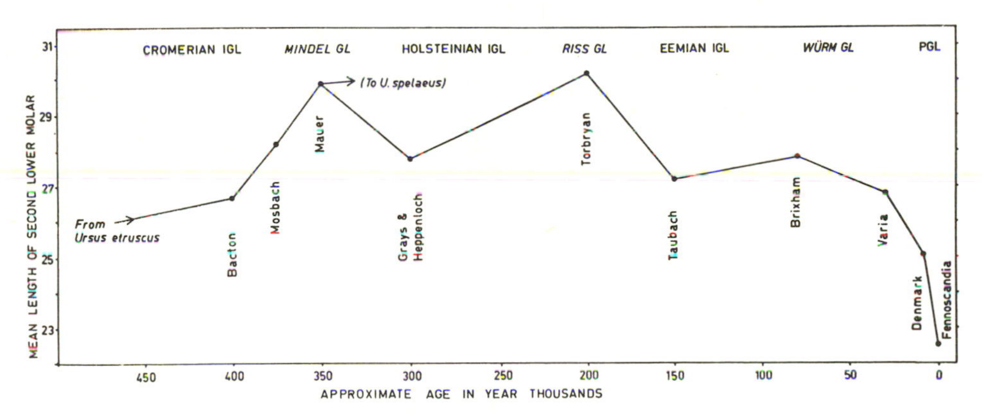

Fig. 8. Size trends in the Pleistocene history of brown bears (*Ursus etruscus* to *Ursus artcos*) in Europe, based on the length of the second lower molar. (From Kurtén 1959.)

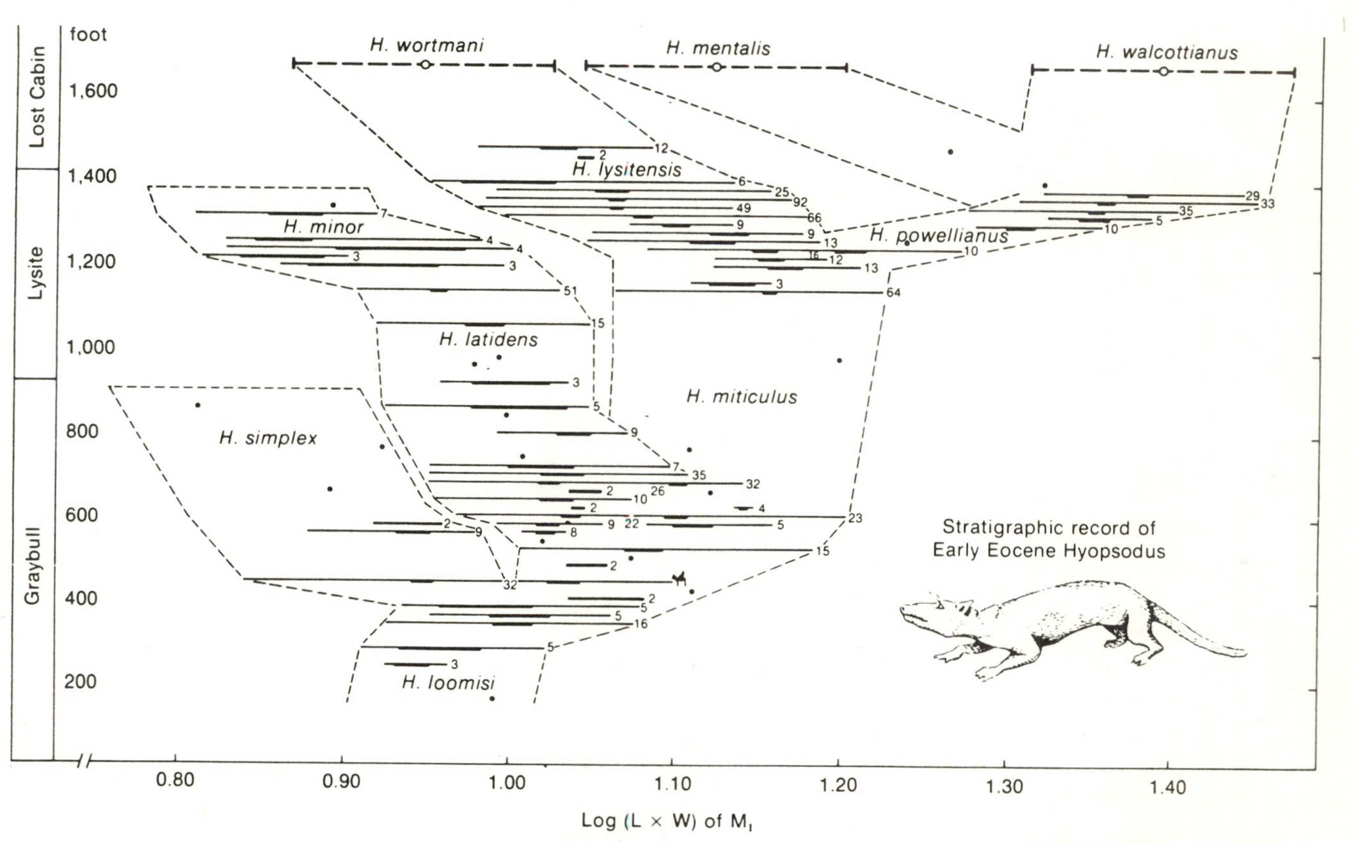

Fig. 9. Stratigraphic occurrence of the condylarth genus *Hyopsodus* in the Early Eocene of northwestern Wyoming. On the horizontal axis is plotted the logarithm of the product of the length and width (in millimeters) of the first lower molar. The heavy bar represents the standard error for a sample, and the light bar its range of values. Numbers to the right of bars indicate sample sizes (number of specimens). Solid dots represent single samples. Open circles and horizontal dashed lines at the top are means and expected ranges of species poorly represented in the stratigraphic section studied but well represented elsewhere. The stratigraphic section represents about 5 million years. (From Gingerich 1974.)

the measurements are made, they show an inverse relation (Fig. 11). In other words, the shorter the time span over which the measurements have been taken, the faster the rate of size change. This is certainly borne out in Figure 7; the steep slopes of various invertebrates and vertebrates measured over a few million years are far greater than the slope of the curve for maximum size which spans billions of years. One of the significant consequences of this beautifully simple graph is that in comparing the rate of size change of any two groups of organisms, it is imperative to take into account the time span over which the measurements were made. As Gingerich points out, many of the comparisons made in the past now need fresh scrutiny in the light of his discovery.

More important to us here is the explanation of the relationship between time span of measurement and size change. Obviously there are relatively rapid increases and decreases in size over short periods of time, and these become averaged out as the time interval of measurement increases. Any trends in size increase must mean that for the time interval measured, the rapid, short-term trends of increase must have exceeded those of decrease. This still does not make clear why there is a trend towards an overall increase in size; why are there so many cases of animals that follow Cope's rule, and why does the maximum size of all organisms increase over geological time (Figs. 5, 6, 7)? At the turn of the century, these trends were considered instances of orthogenesis, which was the term for a progressive trend that somehow, at that time, seemed to satisfy as an explanation. Today it appears to us more mystical than explanatory; we need a straightforward reason for these size increases over great spans of time. The most obvious one is that by becoming larger the organisms enter new size worlds where, among other things, they avoid predation and competition. On the other hand, any sustained selection towards size decrease would lead directly to size worlds of more intense competition, and therefore would be correspondingly rare. But this is a matter we will examine in the next chapter.

There are two consequences of persistent size increase. One is that the anatomical structure of the animal or plant must be able to support the new, larger size. If one took as an example the size increase found in dinosaurs, obviously there would be a limit to how large an animal of this build could manage to exist. All sorts of factors have come into play to make the bones, for instance, disproportionately thick to support the weight, but even these modifications will have a structural limit. Perhaps the most astounding thing is that the basic reptilian anatomy

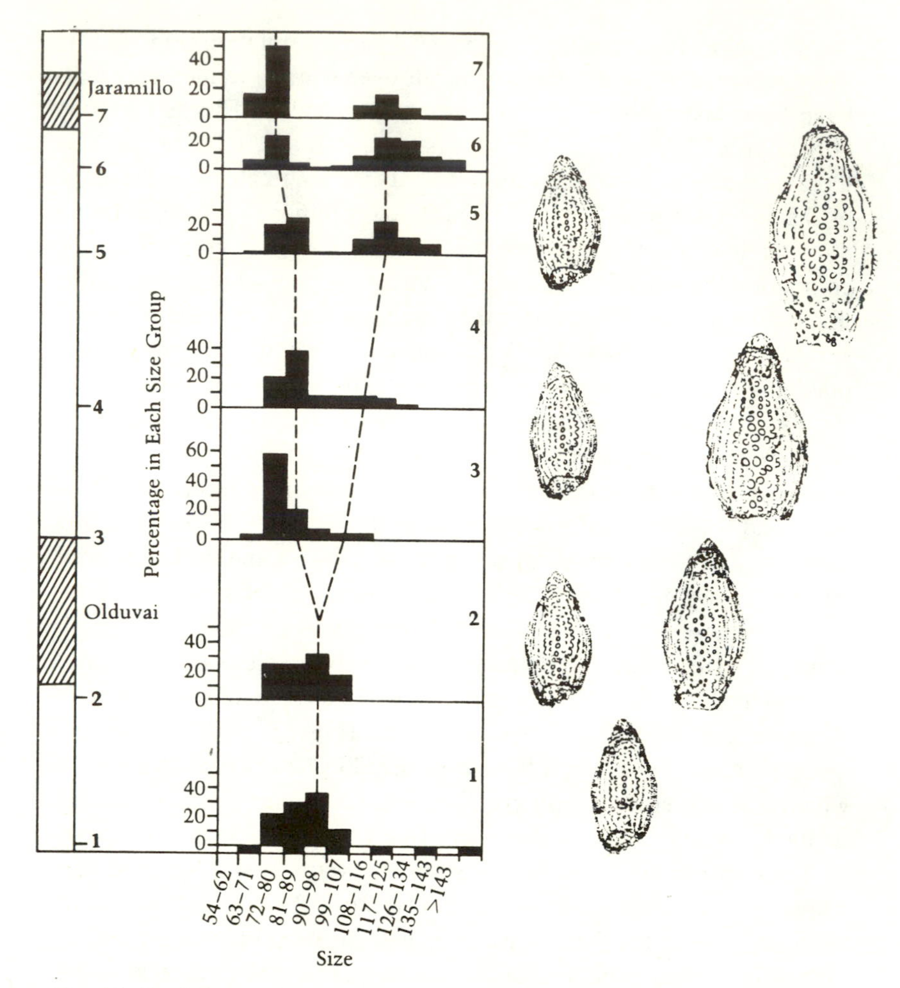

Fig. 10. Rapid but gradual branching of a second species from a persisting ancestor over a span of more than half a million years. The ancestral species is a radiolarian (a small, one-celled organism with a silicaceous skeleton). *Eucyrtidium calvertense*, and the separate branch, is an eventually larger species, *Eucyrtidium matuyamai*. Both are shown here to the right, greatly enlarged. The data are from a sea-floor core in the northern Pacific. The core has been dated paleomagnetically by the paleomagnetic episodes designated as Olduvai and Jaramillo, which are shown in measured positions in the core. The black histograms show the distributions of sizes of *Eucyrtidium* in samples from seven different levels in the core, numbered from the bottom up. The specimens were placed in ten different size groups, which are scaled horizontally at the bottom of the figure. The histograms are scaled by percentage in each sample of each size group.

At the bottom of the figure, both sample 1 and sample 2 indicate a single species, the ancestral *E. calvertense*, with

Fig. 10 Cont.

some (but not great) variation in size. By sample 3, the indicated variation is markedly greater than is usual in single species, but there is as yet no clear indication of a division into two distinct parts. In samples 5 and 6, there is decidedly such a division, although there is still a very small number of intermediate specimens. In sample 7, there are no intermediates, and the two lineages clearly represent two quite distinct species. The average size of one (to the left) has become slightly smaller, but it is essentially stabilized and is still considered to be a surviving ancestral *E. calvertense*. The branch species, *E. matuyamai*, seems also more or less stabilized in samples 5, 6, and 7, but its later history is not followed in this particular core. (From Simpson, *Fossils*, 1983, Copyright © by *Scientific American Books*, based on the work of Hays and later Prothero and Lazarus 1980.)

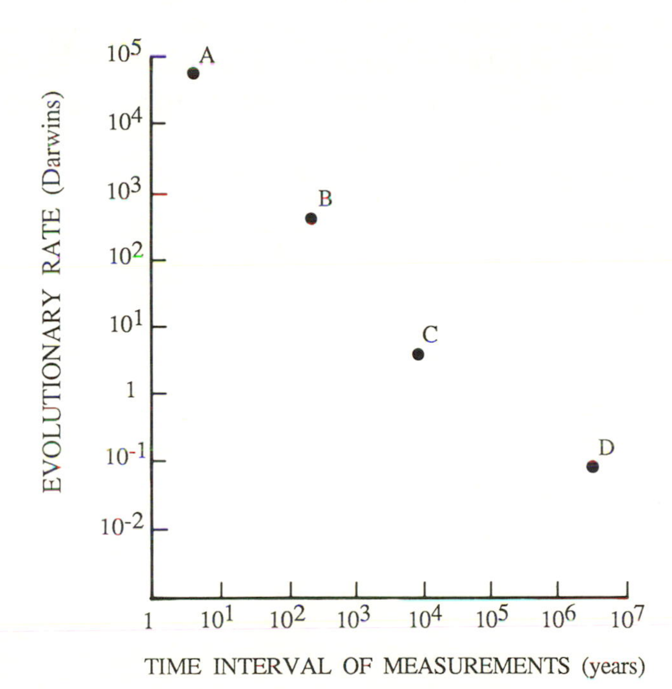

Fig. 11. A graph showing that evolutionary rate is inversely proportional to the interval of time over which rate is measured (on a logarithmic scale). The points indicated are means. *A*: size changes from laboratory selection experiments (sample size, $n = 8$); *B*: size changes in animals recorded in historical times ($n = 104$); *C*: faunal size changes in the Pleistocene following glaciation ($n = 46$); *D*: size changes over long time periods from the fossil record ($n = 363$). Rates are given in darwins. A darwin is equal to slightly more than the doubling of size over a million years. (Actually 2.3 ×, the base of the natural logarithm, e.) (Drawn from data of Gingerich 1983.)

was able to adapt to produce an animal the size of *Brachiosaurus* (Fig. 12). It is quite obvious that a coelenterate (cnidarian), such as a jellyfish, would never be able to produce anything this size, even in water, and certainly not on land, where it would be more affected by the forces of gravity. The largest living animal, the blue whale, can account for a major share of its huge size to the fact that it is aquatic. For this reason *Brachiosaurus* seems to be an even more remarkable feat of biological engineering.

If we now compare the long-term, maximum size changes that alter very slowly over billions of years and the size changes within groups that follow Cope's law, there is an important message. The maximum sizes are each attained from a different, major group of organisms: for plants they are, for instance, lycopods, gymosperms, and angiosperms; for animals they are trilobites, eurypterids, fish, amphibians, dinosaurs, and mammals (Fig. 5). These maximum sizes were reached by progres-

Fig. 12. The skeleton of the *Brachiosaurus*, the largest kind of dinosaur known. It is thought to have weighed more than 85 tons. (From Janensch, *Paleontographica*, Supp., vol. 9, 1929.)

sive size increases within each group. This means then that overall size change during evolution is a result of a series of rises of different successive taxons (Fig. 13).

The same argument applies to Cope's rule. The crashes are rarely due to prolonged periods of size decrease because examples of such trends are so rare in the fossil record. Rather, they are due to extinctions, of which the disappearance of large dinosaurs is the most celebrated example. There is some interesting paleontological evidence of S. M. Stanley (1979) that larger organisms are more likely to become extinct than smaller ones. He has shown that it is possible to measure the average duration of the existence of species of a particular fossil group, and one can make rough estimates of the size of the animals in the group. If Stanley's species longevity data are plotted against approximate size, one can see (with a few notable exceptions) that there is a plainly negative correlation (Fig. 14). He makes the point that the groups with short-lived species are more complex than those with long-lived ones and suggests that elaborate adaptations favor a faster turnover of species. He indicates that Darwin also saw this relation and said, "There is some reason to believe that organisms, considered high in the scale of nature, change more quickly than those that are low; though

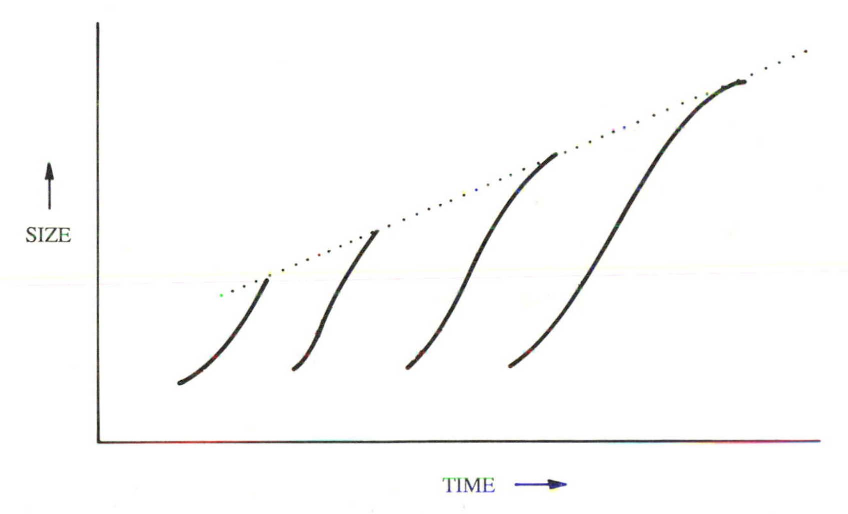

Fig. 13. A hypothetical history of size increases for successive taxons during the course of evolution. The dotted line at the top shows the upper size limits.

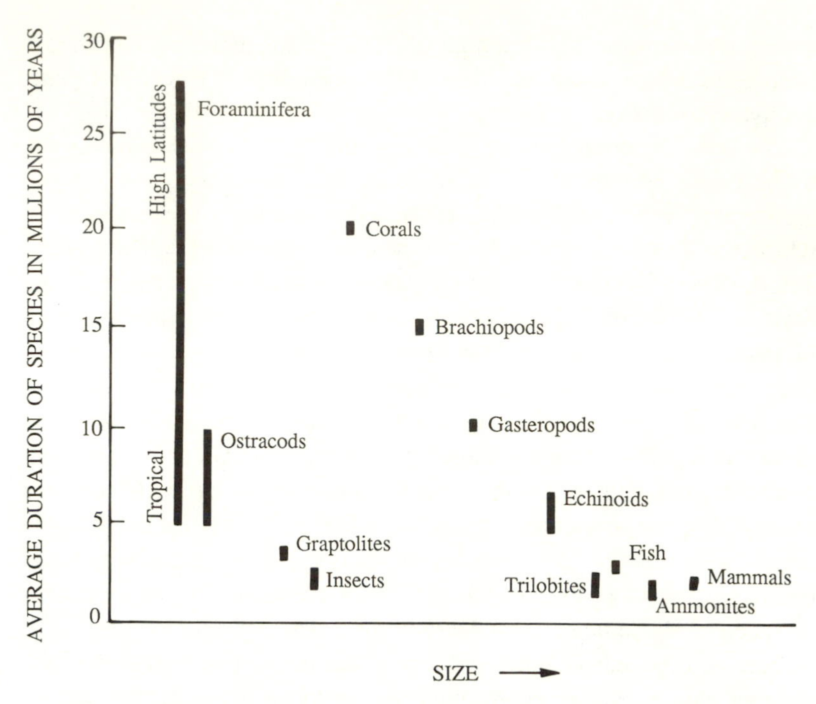

Fig. 14. The average duration of species for different groups of animals (extinction rate) plotted against their approximate relative size. (Data from Stanley 1979 and some help from Gerter Keller.)

there are exceptions to the rule'' (1859:313). Later we will discuss in some detail the relation of size to complexity. Suffice it to say here that the two are positively correlated, and that species that persist for relatively short periods of time, that is, are prone to extinction (as well as the formation of new species), tend to be larger and more complex. The only point that remains obscure is why insects, graptolites, and ostrocods should be such striking exceptions to the rule; perhaps the rule only applies for the upper limits of any one size group.

The fossil record shows repeated evidence for the extinction of large organisms; now, although we anticipate the discussion in the next chapter, we must ask why this is so. The answer undoubtedly lies in the fact that the larger an organism, the greater its requirements become. In the case of an animal, it needs an enormous quantity of food per individual and any serious threat to this supply would mean a devastating loss to a population. Imagine the problem that might face a herd of elephants in

Africa if there were a very prolonged drought. It would mean that all the vegetation available to them would disappear and inevitably the population would eventually die. Contrast this now with some small mouse that also exists on a vegetarian diet. It too will be decimated, but a few individuals will have a greater chance of surviving than the huge elephants; they are so small that they could make do with a scattering of minute plants near some spring. When the rains come back, all the elephants will be gone, but the mice, although only a few individuals are left, will begin to reproduce with extraordinary rapidity and repopulate the green earth. The lesson from this ecologist's Aesop's fable is that the elephant is less pliant in the presence of environmental catastrophe and, in this sense, it is more susceptible to prolonged adverse environmental changes. The same argument could be made for tall trees and minute grasses; it is a general principle. Also note that the larger species will initially be less abundant, thus increasing the probability of their extinction.

To return to the point that major new groups arose from small ancestors, the vascular plants arose as small, erect stems, somewhat similar to the club mosses we find in our deciduous forests today (Fig. 2). From these sprang the huge horsetails, lycopods, tree ferns, gymnosperms, angiosperms, and numerous other groups, all with very large representative species. It is particularly telling that the club mosses that exist today are only a few inches high, yet 300 million years ago there were enormous club moss or lycopod trees 30 meters or more in height. As in our fable of the mouse and the elephant, the large ones became extinct while the small ones survived.

Among animals, an excellent example of the rise of a new group from small ancestors is the evolution of mammals (Fig. 3). They arose from reptiles during the very period when huge dinosaurs were flourishing, and they survived the extinction of the dinosaurs during the Cretaceous. Again it is possible to assume that their ability to survive was in large measure due to their small size. There is always a raging controversy over what factors were primarily responsible for the extinction of the dinosaurs. That it was due to some sort of sustained or abrupt change in environmental conditions are two of the favored ideas and both fit in with the hypothesis put forward here.

In these examples there is another message that is slightly hidden. When vascular tissue in plants, or the anatomical changes of mammals, evolved, they increased the efficiency of small plants and animals, and no doubt, as J. S. Levinton (1986) suggests, these innovations arose over a long period of time involving many sequential steps. That they could at a later time in evolution support larger organisms is entirely fortuitous. In any event, the upward surges on our size-time curve (Fig. 13) are the result of new inventions in anatomy at the low-size points in the evolutionary curve, and these inventions are often successful in supporting huge organisms as a response to selection for size increase.

What we have done thus far is to give the broadest possible picture of size in evolution. We have been operating on a grand size scale, but fully appreciate the fact that in order to really understand the relation of size to evolution we must examine the problem at a finer-grained level. This means we must look at the distribution of size of animals and plants in nature, and try to understand the details of how natural selection operates within the time span of generations.

Further References

For a good, extensive discussion of all the consequences of aclonal versus clonal life cycles, see J.B.C. Jackson, L. W. Buss, and R. E. Cook (1985). For a useful discussion of Cope's law, see S. M. Stanley (1973).

Chapter 3

The Size of Organisms in Ecological Communities

Size levels in nature ▪ The relation of organism size to their abundance ▪ Size and life history properties ▪ Size changes within a species ▪ Size in sexual selection ▪ Conclusion: There is always room at the top

THE PURPOSE of this chapter is twofold. One is to bring together some of the ideas in modern ecology that relate to the size of organisms and the role of size in community structure. But much more important to the central theme of this book is the search for an ecological explanation of size change that would apply to the major changes we have just discussed that occur during the course of evolution. It is the quest for a reasonable microexplanation; we want to see how changes that have occurred over short time intervals could ultimately lead to larger changes that can only be measured over geological time spans.

We will begin with a description of size levels that exist in the living communities that surround us, and then attempt to understand these levels in terms of interactions between and within size levels.

SIZE LEVELS IN NATURE

Size has been a matter of considerable interest to ecologists for many years. Their concern has stemmed largely from energy considerations: what and how much do organisms eat? This gave rise to the concept of food chains and other matters that bear on the relations between prey and predator, including C. S. Elton's (1927) idea of food pyramids. In any one animal community there are first the basic herbivores that graze and consume the vegetation. These are preyed upon by carnivores that, in turn, are preyed upon by larger carnivores. In this way Elton produced a pyramid that can be described either in terms of how much energy is available at each level or on the basis of the size and number of animals at each level (Fig. 15).

It was clear from the beginning that the individual size of the organisms at each level could vary enormously. For instance, Elton pointed out that a successful carnivore could attribute its success to large size, although in some instances small carnivores could make up for their slight build by being exceedingly aggressive. Another way of allowing the predator to be smaller than the prey is cooperative hunting. Wolf packs can bring down a huge moose, or hyena packs can easily fell a large wildebeest.

In food webs the size of both the prey and the predator show great variation. For instance, some large toothed whales such as the sperm whale eat huge squid and killer whales eat seals and even sometimes, in a group, will attack a whale, while the huge baleen whales eat krill, a small shrimp that exists in great quantities in polar waters. Such facts are obvious and have been recognized for a very long time. Because the exceptions to any kind of rigid size sequence between prey and predator have captured the main interest, food chains and food pyramids, although important for other ecological considerations, have not been very helpful in giving us an insight into size relations in nature. For this we must turn to new directions.

The Relation of Organism Size to Their Abundance (Density)

First let us ask what is the size distribution of animals and plants in any natural habitat. Take, for example, a wooded area in a temperate zone such as the eastern United States. In the soil one will find great masses of bacteria which constitute one end of the scale. At the other end there will be large mammals such as deer (or moose and bear in the northeast) and large trees such as beech, or tulip trees, or white pines. This range of size extremes is at least seven orders of magnitude: from micrometers to meters, if one measures in lengths. Note that this size span is similar to the one we encountered in the whole span of evolution (Fig. 5), and it was pointed out in that discussion that small and intermediate-sized organisms were retained during the course of evolution. Individual species and whole groups may go extinct, but it is obvious that in our woods all the intermediate sizes are represented.

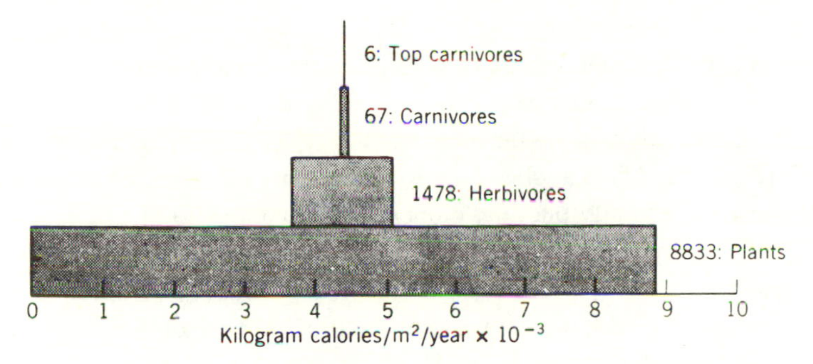

Fig. 15. The "pyramid" of productivity in a community. (Silver Springs, Florida.) (After Odum, *Ecol. Monographs*, 27:55–112, 1957.)

It is possible to identify the general characters of these intermediate forms: the size above bacteria would be amoebae that live on bacteria; they might be considered the main grazers in the soil. They in turn are eaten by a variety of small invertebrates, some of which eat bacteria also. These would include some large ciliate protozoa, rotifers, flatworms, and nematodes, the latter being especially common in soil. These small invertebrates will be preyed upon by larger ones, including mites and the larvae of many insects. We now reach a size level over a centimeter in length, and this would include earthworms, amphibians (such as newts), small rodents, and even small birds. However, many vertebrates will be larger than 10 centimeters in length, including numerous birds, some amphibians (such as frogs and toads), reptiles, and mammals such as raccoons, rabbits, woodchucks, bobcats, and so forth. Finally, the largest animals will be deer, moose, bears, and possibly a mountain lion. If one did the same kind of progression with plants, at a lower end of the scale, following bacteria, there would be small algae and fungi. Somewhat intermediate in size are lichens, mosses, and liverworts; and then there are angiosperms, which range from the smallest grasses, to flowers, to shrubs, and finally to trees, small and large. It is interesting that gymnosperms have very few small forms; they exist almost entirely as trees, such as pine, fir, and spruce. Presumably small angiosperms have outcompeted the more primitive, small gymnosperms.

The next step in understanding the size distribution in these woods is to look into the relation between size and population density. This has been reviewed for animals by R. H. Peters (1983), and the general conclusion is that animal density (the number of individuals per square meter, or kilometer) is inversely proportional to the size of the animal; that is, small animals will be far more numerous in a given area than large ones (Fig. 16). (This applies only to aclonal organisms and not to large colonies of asexually budding clones.) The exact ratio of this relationship is not quite so clear. There is not only variation between taxonomic groups, but possibly between different communities (J. H. Brown and B. A. Maurer 1986). This inverse relation between size and abundance has also been demonstrated by D. R. Morse et al. (1985) in a novel fashion using fractals to analyze insect populations in various tropical communities, and their results are in a general way consistent with the overall conclusion summarized by Peters (Fig. 16).

Another way in which the relation between size and energy considerations can be shown is to examine the relationship between the size of an animal and its home range, the latter being a reflection of the space needed for food gathering. This has been examined by numerous authors, and indeed body mass is roughly directly proportional to the size of the home range: the bigger the animal, the larger its home range.

But in energy considerations, the most important point is that the

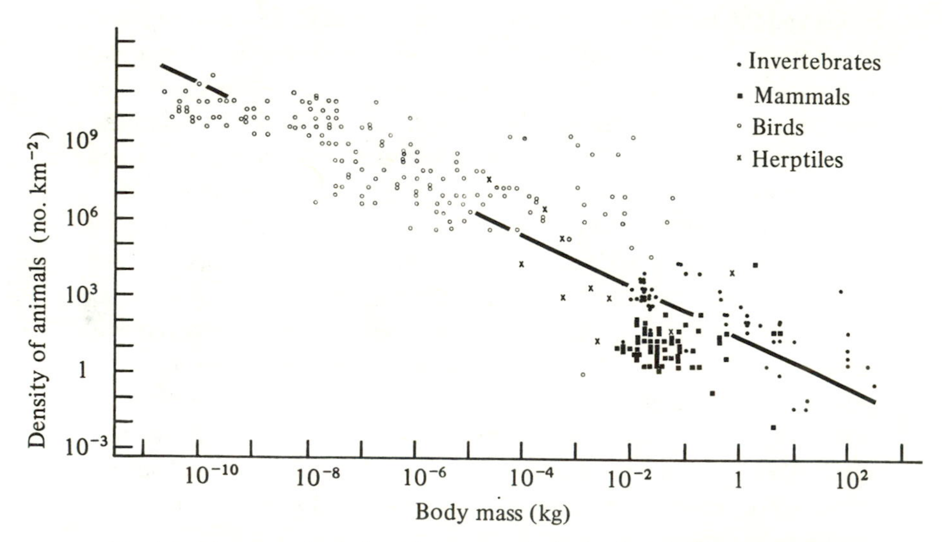

Fig. 16. The general relationship between the body size of different animals and their population density (abundance). (From Peters 1983, Copyright © by Cambridge University Press.)

higher size levels depend on the lower ones for food. At least this is clearly the case for animals; the larger animals could not exist were it not for the smaller ones (as well as small plants for many herbivores). We have come right back to Elton's food pyramid and conclude that one of the reasons that we have an enormous size span of animals in any particular community is that the larger ones, at least the carnivores, are utterly dependent for their existence on the smaller ones as sources of food. Furthermore, the available food at one size level determines the number or density of individuals at the next higher level. Therefore, as far as animals are concerned, the general relationship we see in Figure 16 is rigidly and explicitly laid down by energy considerations. This, in part, may help us to see why, during the course of evolution, when large animals increased progressively over time, all the small ones did not go extinct.

There is abundant evidence that the relation holds for plants too. For instance, in experimental plant populations (i.e., where only one species of plant is grown in a given plot) the average plant weight is inversely proportional to the density of the plant; in other words, it is similar to the line shown in Figure 16. This is even true of growing populations; as the plants increase in size, those with faster growth rates at early stages will outcompete the less-endowed ones so that the latter will be progressively deprived of water and nutrients in the soil as well as the sun as they recede into the shade of the fast growers. The result is a progressive die-off so that the weight-density relation remains constant as the remaining plants increase in size.

O. T. and D. J. Solbrig (1984) point out that the same principles apply in nature where there are many different species of plants competing with one another. Ultimately, the large trees win by expanding their leaves over the smaller competitors of their own species, and those of other species as well. The smaller species win by other devices, such as tolerating shade or making use of small patches of sun that escape through the larger trees. It must be remembered that a plant community, such as a forest, is not a permanently fixed structure, but one that constantly goes through succession from small shrubs to the larger trees of the climax forest. For even in the latter, large trees will die and leave a gap, or the forest will come to the edge of the meandering river where new bank is constantly being accumulated on one side and old forest gouged out on the other, and all these denuded zones, no matter how small, begin the sequence of succession. The important point is that no matter where any plot of land is in its succession, it will reflect the size-density relation. Therefore, it is clear that size and density in plants are related in the same way as among animals, and the reason in both cases has to do with energy considerations. The difference between photosynthetic plants and animals is that the way they capture energy is radically different: sun-catching versus eating.

This brings us finally to the matter of different environments. Size relations and abundance of both animals and plants will be, as is self-evident, significantly affected by climate and other environmental circumstances. One would not expect to find the same values for a desert or tundra as for a temperate forest. And there are density differences between tropical and temperate forests. These have been a major concern to ecologists for many years. Of special interest to them has been the relation between habitat and species diversity, or the number of species—a matter which will be discussed later when we get to the subject of ecological complexity (Chapter 5).

Here our main interest is not in diversity or in energy relations, but in the size structure of the community. First, I will review briefly what is known concerning the relation of the size of organisms (from the smallest to the largest) to properties of their respective life histories. With this as a background, we will examine size change at one size level, within one species or between closely related species. We can then see if these recent, relatively rapid size changes provide a basis for an explanation of the grand size changes we observe over millions of years of evolution.

Size and Life History Properties

The most fundamental consequence of size differences on the life history of organisms is their effect on reproduction. In brief, if the organism is small, the generations are short and the reproductive rate is high.

Size is correlated with generation time in a very clear-cut way (Fig. 17). The basis of this relation is very straightforward: smaller organisms have a shorter period of development, simply because it takes them less time to build. Since, in general, organisms do not produce egg or sperm, or reproduce asexually until they have reached maturity, it follows that the larger the organism, the longer it will take to achieve its first reproduction. I have discussed these arguments in detail elsewhere (Bonner 1965, 1974) and will not dwell on them here.

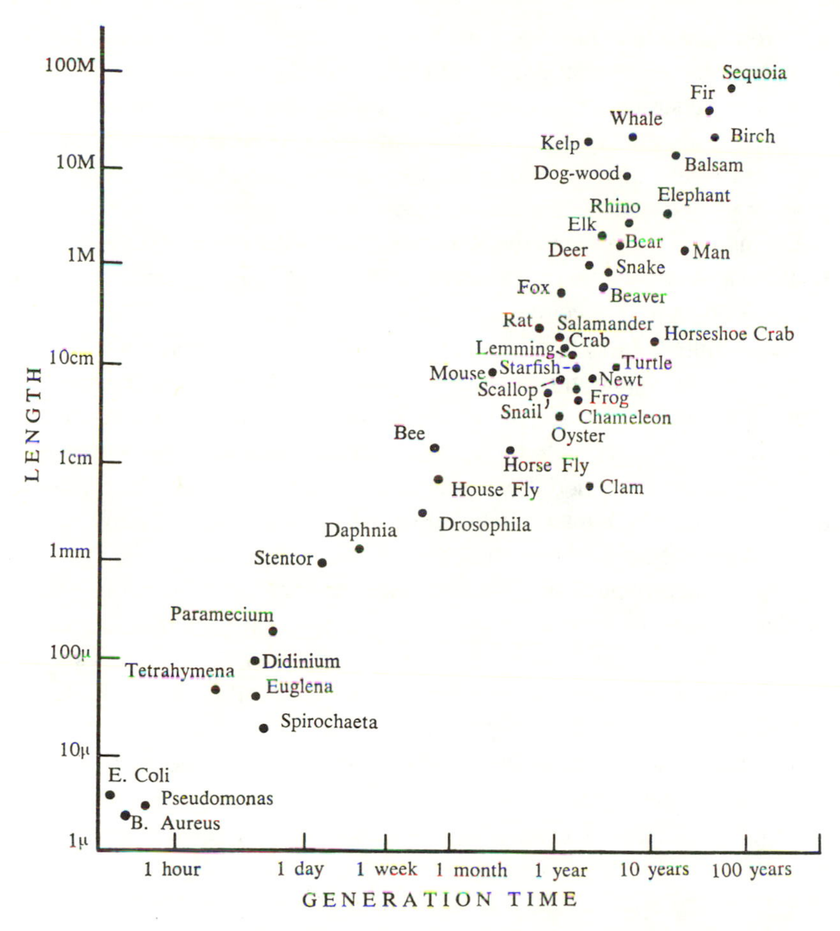

Fig. 17. The length of an organism at the time of reproduction, plotted against its generation time on a logarithmic scale. (Bonner 1965, Copyright © by Princeton University Press.)

It should be pointed out that this relation applies only to the first reproduction. Many large animals and especially plants, such as trees, will continue to produce offspring for many years after the first reproduction, and in numerous cases they continue to grow in size. This leads me to an aside: many large trees do not set seed until they have reached a considerable size (Fig. 17). For instance, giant sequoias do not reproduce until they are 80 meters tall, which takes 60 years to achieve. This seems strange because there is no physiological reason why they could not propagate when they are a few feet tall. One can only guess at possible answers to this restraint on their part. Putting energy resources into reproductive structures is something a sapling can ill afford when it is struggling desperately to grow more rapidly than other rival saplings. Only the fastest growers will win in the competition for sun, and any plant that diverts its precious resources towards cones or flowers and seeds may lose. One hypothesis is that giant sequoias wait so long to fruit because this is the best adaptive strategy; it is an argument which depends on natural selection.

As one might expect, in aclonal organisms size also correlates with longevity; the bigger an animal, the longer it is likely to live. It is also true for mammals that the larger the species, the longer the gestation period. A particularly interesting effect of size and generation time is found in mammalian herbivore population cycles. Recently, R. O. Peterson et al. (1984) made an extensive survey of the literature and showed that the larger the animal, the greater the time span of the cycles in a fluctuating population (Fig. 18). So there are many consequences of size on life history events, but note that the examples in this paragraph primarily apply to vertebrates, especially mammals. This is simply because they have been most extensively studied; no doubt similar interesting size-dependent properties will be found for all organisms.

The main purpose of our present discussion is to examine the relation between size and reproduction, and this brings us directly into the realm of *r* and *K* selection. The basic idea is very simple: organisms have two alternative strategies to produce offspring for successful reproduction. One is to have a rapid increase in the number of offspring (*r* selection) and the other is to have few offspring, but those that are produced are protected and cared for in various ways to increase their chances of survival (*K* selection). These two strategies are correlated in a general fashion with a number of characteristics: *r* selection, with its high, net reproductive rate, is characteristic of small organisms with short generation times, while *K* selection favors the converse. This is a continuum and it serves as a useful way to distinguish differences and compare two groups of organisms.

What is especially important is how these differences relate to environmental stability. A *K*-selected organism will tend to be favored in highly diverse, stable environments, while *r*-selected ones will manage better in fluctuating, unstable environments. In the former, there are few new opportunities for the young, and therefore each one is carefully

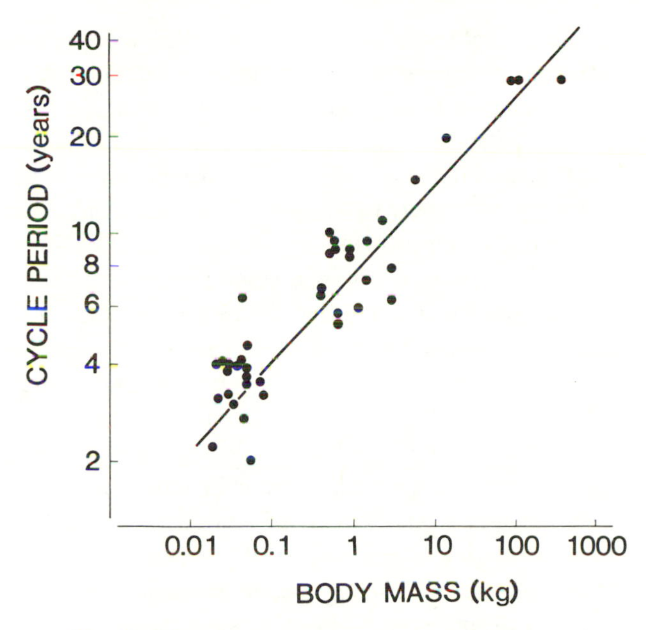

Fig. 18. The relation between the duration (period) of population cycles and body mass, shown on a logarithmic scale. (From Peterson et al. 1984.)

prepared so that it can find its place amid intense competition; while in the latter, a whole generation may be wiped out as a result of sudden environmental shifts, and the best means of regaining a position in the community is by flooding it with numerous offspring. Much has been written about *r* and *K* selection, criticizing it in detail and citing all sorts of problems; however, I must emphasize here that I am only using the concept as a helpful but very rough generalization.

Its usefulness to our argument stems from the fact that numerous ecological properties correlate reasonably well with organism size. So if we consider the animals and plants in Figure 17 (which is a continuum), then the smallest bacteria will indeed be the supreme *r* strategists; they have a very high reproductive rate and live in extremely perilous and frequently fluctuating environments. At the other end of the scale, huge mammals and trees—the top *K* strategists—become greatly specialized and manage to exist in a stable state for extraordinarily long periods of time. They manage this in two ways: one is that by being large, they simply override the minor perturbations that unicellular organisms suffer. Elephants and giant sequoias exist smoothly through periods of extensive rainfall, long periods of sun, or larger fluctuations of heat and cold; their body physiology is resistant to such minor changes. The second way in which large mammals maintain stability is by having long periods of gestation, offspring care, and homeothermy, all of which isolate them from environmental dangers in a dramatic way, something quite impossible in the microworld.

With great increase in size comes a lack of flexibility, a degree of specialization. Large size successfully isolates the organism from minor environmental changes, but by the same token it is especially susceptible to extinction in major climatic or environmental shifts. A large *K*-selected organism cannot recover quickly from decimation by catastrophe, for its reproductive rate is so low, while a minute *r*-selected beast can inundate the environment with offspring the moment conditions are favorable. We discussed this difference between large and small in Chapter 2 (our fable of the mouse and the elephant), and it may account for the fact that extinctions often occurred in the geological past in large species, while new groups stem from small ancestors.

The discussion so far has been purposely kept to large size differences. This has been helpful to understand the structure of a community and why, during the course of evolution, there has been a maintenance of organisms at all size levels. It also helps us to see that some of the major events of evolution—the formation of new groups and the extinctions of others—may be at least partially accounted for by properties of the organism that are related to size.

SIZE CHANGES WITHIN A SPECIES

Earlier I said that a microexplanation of evolutionary size change is being sought, and this must be at the population level. There has been much argument recently as to whether the standard views of modern population genetics are sufficient to account for major or macroevolutionary changes, or whether new principles are needed. In contrast to some of the modern paleontologists, who include N. Eldredge, S. J. Gould, and S. M. Stanley, many biologists continue to find no difficulty in applying principles of population genetics to all evolutionary change.

Part of the difficulty is that what a paleontologist considers a species may be a quite different thing from the species of a biologist. Even the biological definition is an enormous problem and may well be different things in different groups of animals and plants. It is clear, if we look at larger populations of similar organisms living in nature, that there are discontinuities; and sometimes those discontinuities are sufficiently sharp so that these populations are called species, but at other times they may be less distinct, and therefore the populations are called subspecies, or varieties, or races. There is currently an enormous, complex discussion as to how to judge species, and how to find appropriate criteria; but the furor is not just a recent phenomenon, it is an ancient one. It involves questions as to how best to judge differences between organisms and what are the most meaningful guides. Unfortunately, it is difficult to find any one set of criteria that is satisfactory for different groups of organisms, and sometimes it is equally difficult to find a perfect set for a small restricted group.

The reason for this is probably explained by the fact that we are seeing, among populations that differ to varying degrees, the incipient steps of species formation. For some reason, groups of individuals may have become isolated in the sense that they no longer interbreed with neighboring groups and, depending upon the degree and duration of the isolation, the populations will diverge to different extents in their genetic constitutions and morphologies.

The way in which populations show differences does to a great extent depend on a number of characteristics, and this is the main reason for saying that the criteria for species may differ with different groups of animals and plants. For instance, in the constitution of the genome there may be differences in the number of chromosomes, their ability to form extensive chromosome arrangements without adverse effects, the size of the genome, and many other of its properties. The same is true of morphology: different organisms will differ in the way they can produce slight morphological changes that result in the formation of a new species. What is especially interesting to us here is that the ability to modify morphology is size-related. This is a matter I will touch on briefly here but examine in detail in Chapter 4. The larger the organism, the longer and more complex is its period of development. This means that the way in which the adult morphology is determined varies; the road from the transcription of a gene to the final morphological effect may differ enormously between large and small organisms. To give a crude example, a unicellular organism can, by mutation, easily make a basic change in the structure of some protein that will affect its cell shape, for instance change it from a cylinder to a sphere. Any mutational change in a vertebrate or an angiosperm can affect only some minor character, such as fur pigmentation or leaf indentation; changes in the whole structure of the large organism would either be impossible in a way comparable to that of microbes, or it would take many millions of years.

To return to the main thread of my argument, here we are concerned with a microexplanation of the evolution of size. Size is like any of the morphological features that might change during evolution; we have simply isolated one example. Because it is so easy to measure, there have been many studies of the size change among living populations, and there is a large literature on the subject. Here I will bring together some of the main conclusions.

It is simple to demonstrate that a size change has occurred in a population; it is far more difficult to explain the reasons for the change. There have been numerous studies comparing a species of an island with its parent stock on the mainland. After many years of isolation the island race will often be larger or smaller than the mainland form; it is relatively rare that they remain the same size. To cite some well-known examples, the extinct elephants of Malta were dwarfs, and the present-day deer of the Florida Keys are diminutive, while many different species of rats, moles, and mice have larger island forms. There has been a great deal of speculation about what could be the causes of these trends. In general, various authors agree that one important factor is the reduction of the resources on small islands; if there is less to eat, small size will be adaptive. Other factors that might play a part are a reduction or disappearance of a predator. In the presence of predators, some prey increase in size to avoid capture, and others may become small to hide more effectively, but on an island these pressures will disappear. There is also an effect of an island's size on competition between members of the same species; the larger the island, the more intense the competition. But whether these factors should invariably or only occasionally produce an increase or a decrease in size is uncertain. In addition, there is the possibility that some of the size differences may be due to chance, for islands have small populations, which is an ideal situation for the stochastic "genetic drift" of Sewell Wright.

Whatever the conjectured reasons for these size changes, M. V. Lomolino (1985) has provided a major new insight into the problem. He scoured the literature for data on size differences between island and mainland forms of mammals and plotted them in such a way that the change (increase or decrease) in size on the island is compared with its size on the mainland (Fig. 19). From this we can see that, on the average, smaller animals tend to increase in size on islands while larger animals become dwarf. There are many exceptions, but the trend is clear. Therefore, even though the detailed reasons for size change on islands remain highly conjectural, Lomolino's generalization remains clear-cut.

Of the variety of ecological factors which might affect size differences, interspecific competition has been given considerable attention. If two species were competing for similar resources they would diverge in size by selection, so that they could avoid competition: one concentrating on the larger food and the other on the smaller. (This is known as "character displacement.") G. E. Hutchinson (1959) pointed out that often the ratio of the size difference was somewhere in the neighborhood of 2 on the basis of weight (or $\sqrt[3]{2} = 1.26$ on the basis of length). A classic example of the principle is found in bill sizes in Darwin's finches on the Galapagos Islands. D. Lack (1947) showed that if two species were on separate islands, their beak lengths tended to differ sig-

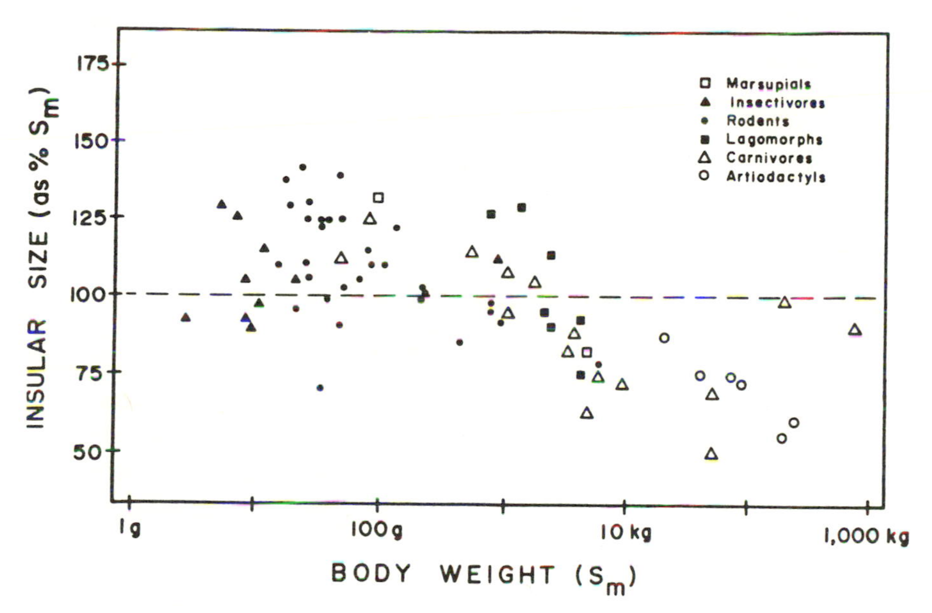

Fig. 19. The ratio of the size of an island form and its closest relative on the mainland (i.e., insular size) is plotted against the body weight of the mainland form (S_m) for a number of different terrestrial mammals. (From Lomolino 1985.)

nificantly less than if they coexisted on the same island (Fig. 20). That this was indeed due to competition was shown by P. R. Grant and his colleagues (D. Schluter et al. 1985). Note that in these cases, and in other similar ones, the competition causes one species to become larger and the other smaller so that the difference between them is accentuated. Therefore, competition can push up as well as down on the size scale. It also has the effect of making the size distribution of related species quite neatly spaced out to conform roughly to the Hutchinson ratio. This spacing is seen beautifully in the fruit pigeons of the rain forest of New Guinea which have been studied by J. M. Diamond (1973) (Fig. 21). There are eight size classes, ranging from small birds of 49 grams to large ones of 802 grams.

The important thing to emphasize from these studies is that size increases and size decreases could theoretically be explained as the result of natural selection. The fact that the selective forces are complex and often working in opposite directions is secondary.

So far we have emphasized the case for islands, but size changes, that is size differences between genetic races of species, are also well known for continental animals and plants. Some of these are associated with climate, a subject that began with C. Bergmann in the last century. It is now well established that the higher the latitude, the larger the individuals of many species of birds, mammals, and even reptiles and amphibians. The reasons for Bergmann's rule are harder to understand. There are convincing arguments why we must doubt that it has to do simply with less heat loss in the larger animals as was originally assumed; this has been effectively refuted by P. F. Scholander (1955). To further complicate the issue, F. C. James (1970) found that size in certain birds was not correlated with temperature gradients, but with a combination of temperature and humidity. In all these instances of climate-correlated size changes, we simply do not understand what the selective forces might be, or even if the size changes are due to selection at all.

Size clines in plants have been known and well studied for many years. For instance, in the classic study of western plants by J. Claussen and his colleagues (review: L. G. Stebbins, 1950), the heights of the plants were inversely related to altitude, and furthermore, this was a genetically controlled size difference, for it would appear even when they cultivated them all together in one garden. In the case of plants,

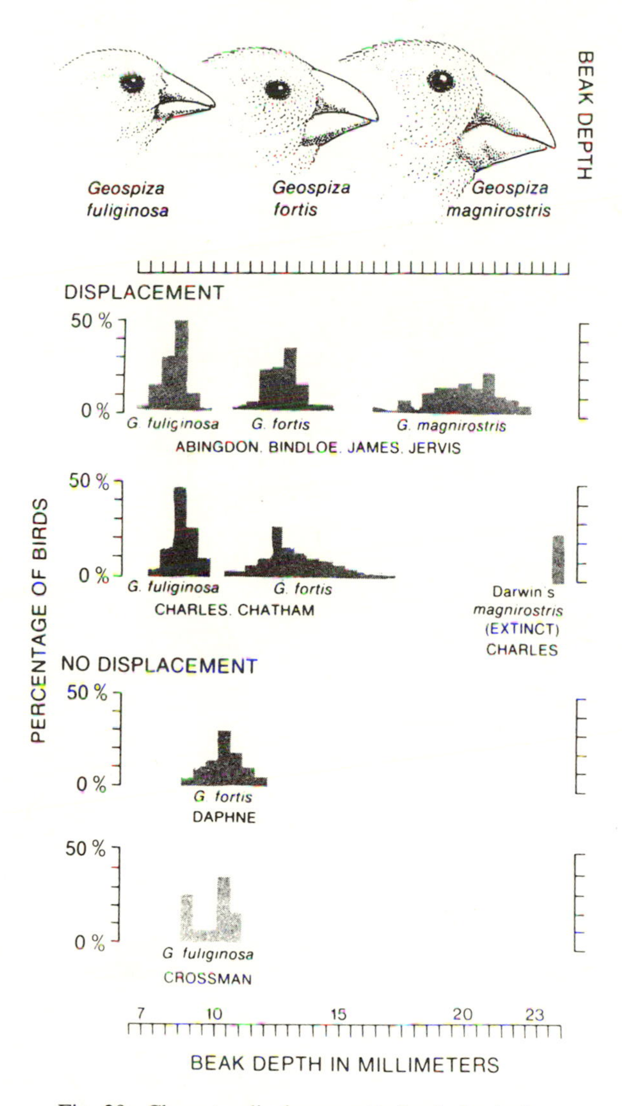

Fig. 20. Character displacement in beak size in three species of Darwin's finches. Note that when species coexist on the same island, their beaks diverge more in size than when they are on different islands. (From Futuyma, *Evolutionary Biology*, 1979, after Lack 1947, Copyright © by Sinauer Associates.)

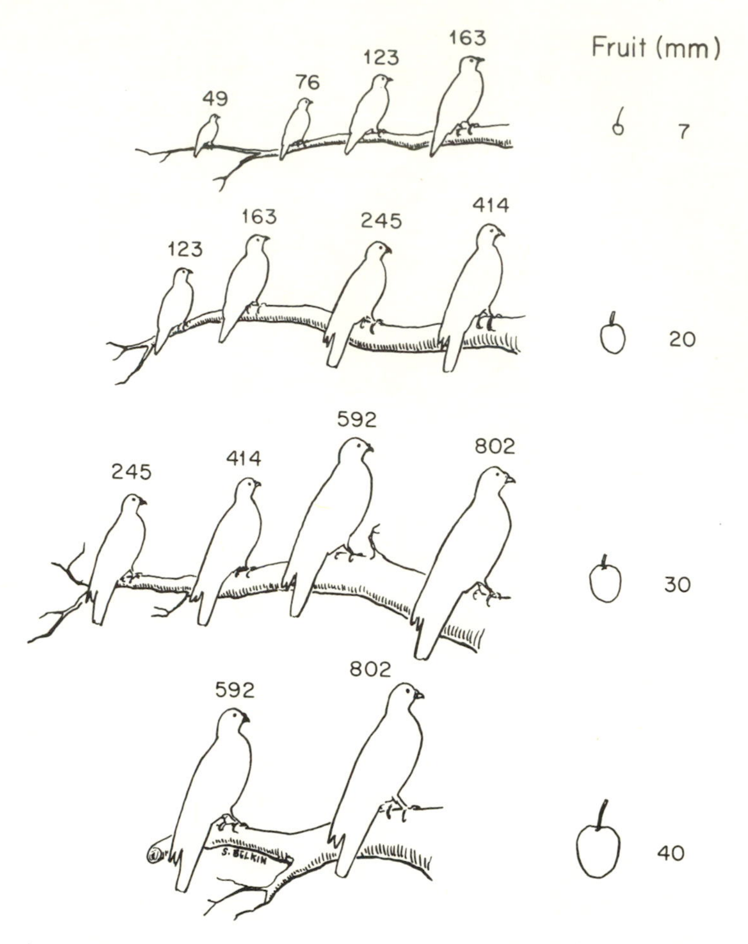

Fig. 21. Schematic representation of niche relations among the eight species of fruit pigeons of the genera *Ptilinopus* and *Ducula* in the lowland rain forest of New Guinea. On the *right* is a fruit of a certain diameter (in millimeters), and on the *left* are pigeons of different masses (in grams) arranged along a branch. The mass of each pigeon is approximately 1.5 times the mass of the next. Each fruit tree attracts up to four consecutive members of this size sequence. Different species of trees with increasingly large fruits attract increasingly larger pigeons. In a given tree, the smaller pigeons are preferentially distributed on the smaller, more peripheral branches. The pigeons having the masses indicated are: 49 grams, *Ptilinopus nanus*; 76 grams, *P. pulchellus*; 123 grams, *P. superbus*; 163 grams, *P. ornatus*; 245 grams, *P. perlatus*; 414 grams, *Ducula rufigaster*; 592 grams, *D. zoeae*; and 802 grams, *D. pinon*. (From Diamond 1973.)

the selective advantages of small size in harsh, alpine environments is easier to grasp, for it is plain to see how smaller plants would fare better under these circumstances. We may not understand the details of how selection operates to produce these size differences, but we can hardly doubt that selection has played the key role.

Size in sexual selection

One of the most interesting instances of size difference within a species comes from sexual selection. Because of either competition between males or choice on the part of females there is often a marked difference between the size of the sexes in many vertebrates and invertebrates. Usually the male is larger, and in such extreme cases as the Alaskan fur seal, the male will weigh 500 pounds and the female only 70 pounds. An equally extreme example of the reverse can be found among invertebrates when the male is a minute parasite of the huge egg-laying female. M. T. Ghiselin (1974) has suggested that such small males provide a mechanism to facilitate dispersal so that outbreeding is favored. Because of their very smallness, the males can be carried considerable distances by currents and attach to distant females, a phenomenon analogous to the strategy of pollen in higher plants.

The importance of these size differences between the sexes, found in so many species, is that it illustrates that size differences, even very large ones, can be contained within the genetic complement of a single species. And furthermore, these size variations between the sexes have arisen by selection. In a dramatic way it underlines the fact that one way to change individual size is through natural selection.

Before leaving the subject of size differences in sexual selection, I would like to pause briefly to discuss an unconventional aspect of that subject. Since selection can act on all phases of the life cycle as well as the adult, there is another point in the life cycle where differences in size between the sexes are great: at fertilization, when the sperm is minute in comparision to the size of the egg. To varying degrees this is true for all higher organisms, the exceptions being only among the more primitive algae and fungi and the protozoa. Again we must ask why this is so, and P. A. Cox and J. A. Sethian (1985) have provided an ingenious model. Numerous authors have considered this subject in the past, but Cox and Sethian add a new way of thinking about how gametes of the opposite sexes find each other. They examine the success of fertilization mathematically in terms of the encounter efficiency and how this efficiency is affected by the size of the gametes.

Small zygotes are associated with small gametes, and Cox and Sethian show that for maximum collision between the sexes, identical, small motile gametes are the most efficient (isogamy). But with size increase of the zygote, it becomes more efficient for optimum encounters if one gamete is very much larger than the other (anisogamy). With this size increase, they argue, the single flagellum will no longer be effective in moving the gamete and a size is reached where for optimum encounters the larger gamete can remain immobile to become an egg (oogamy). The next step for ensuring fertilization has occurred in some organisms: the evolution of chemotaxis, whereby the sperm is oriented towards the egg. Cox and Sethian's arguments are all based on purely chemical and physical considerations that would affect optimum encounter between gametes as their size increases. Many arguments have been brought forward by others for selection pressure for increased zygotic size, which means selection for larger gametes. The most obvious argument is that with an increase in the complexity of the development there are straightforward advantages to a better start in terms of maternal messages in the form of mRNA's, which direct the synthesis of specific proteins in the early embryo, and in terms of accumulated energy reserves in the form of yolk. Therefore, in the evolution of fertilization there has been a size increase in a system in which cells have to fuse in complementary pairs, and the inevitable result has been anisogamy, oogamy, and even chemotaxis of the sperm towards the egg. Again, these are genetically determined differences between the sexes that have risen by natural selection, within the constraints of the microworld of cells.

Conclusion: There Is Always Room at the Top

So far in this discussion I have pointed out that sometimes there is evidence that natural selection is instrumental in producing size differences between closely related species (and even within a species in sexual selection), and in other cases the evidence is wanting. In those latter instances it is equally plausible that selection is not involved at all, and that the size changes are entirely due to random, stochastic shifts. The examples of size changes on islands are especially relevant, because islands, by their very nature, would isolate small populations, and therefore would be in an ideal state for "genetic drift."

It will, of course, be impossible to distinguish in many cases between drift and selection. As interesting as this distinction may be for any particular instance, for our purposes either mechanism is satisfactory, even if selection and drift combine to produce a size change. All we need for our argument is evidence that size does change over time, and we have seen that not only is this true, but there is evidence for roughly an equal amount of size decrease as size increase.

To relate this to what was said in the previous chapter concerning size change over the entire span of evolution, we must remember the important generalization of Gingerich (1983) that the larger the period over which the size change is measured, the slower the rate of the change. The most plausible explanation of this generalization is that size does not go in one direction only, but up and down; therefore, the longer the time span of measurement, the more we can average the increases and decreases and make the total or mean rate of change small. This was well illustrated by the case of Kurtén's bears, as discussed previously (Fig. 8). But on the basis of this argument alone we would not expect any size change at all over extended periods of time; if the short-term increases were equal to the short-term decreases, how could there be a net gain when measured over millions of years?

The solution to this last difficulty comes from the simple fact that in any ecological setting, since all the intermediate sizes are represented at all times, as we have discussed, the only place for expansion, for pioneering, is in the upper end of the size scale. This is a place where totally new species can appear, either by selection or by drift, that will not meet stiff competition. They are pioneers that have avoided competition by finding a new niche that is above all the others. Elephants and giant sequoias are obvious examples. Therefore the very slow progress of maximum size increase over the entire course of evolution, drawn so crudely in Figure 5, represents the rate at which new, upper size limit niches have been invaded over the last 3 billion years or so. While the size changes of any one species may have been produced by random drift, or natural selection, the question of whether the new large races persist or become extinct must be greatly influenced by selection. Certainly extinctions could be stochastic, but persistence over long periods of a new size morph must involve natural selection.

Further References

There is a vast literature on the topics discussed in this chapter. On the subject of scaling in animals, besides R. H. Peters (1983) mentioned in the text, the recent books of K. Schmidt-Nielsen (1984) and W. A. Calder (1984) are important. The subject of plant density is well reviewed by J. L. Harper (1977). For further reading on *r* and *K* selection, see R. H. MacArthur and E. O. Wilson (1967), E. R. Pianka (1983), H. S. Horn and D. I. Rubenstein (1984).

Chapter 4

A Problem in Developmental Biology: Why and How Larger Plants and Animals Are Built

Ways of becoming multicellular ▪ *Selective forces for multicellularity* ▪ *Why development?* ▪ *Development of support structures* ▪ *Proportions and size* ▪ *Developmental steps and size* ▪ *The legacy of past developments*

IN THIS chapter I willl show that one of the consequences of overall size increase during the course of evolution is an extended period of development; it is a period of extended construction. This will bring us to consider modes of building for increasing size, including the first major step, the origin of multicellularity, and the formation of supporting materials and structures. To introduce the subject, I will review briefly various aspects of this period of size increase that I have discussed more fully elsewhere; they will serve as a general background to the new points made here.

Cell size (and here let us concentrate on cells with nuclei, or eukaryotic cells) is associated with a unit of effective metabolism; it is a small energy-converting machine. There are many good reasons for supporting the idea that the rough constancy of cell size for all organisms bears some relation to these energy considerations. If one thinks of the rates of different chemical processes occurring within the cell, the distances needed for diffusion, the surface boundaries needed for isolating different chemical components of the motor, and so forth, all of these lead to the conclusion that there is an optimal size with sharp upper and lower limits, which is the size found in nature. This is equally true of multinucleate or coenocytic organisms such as algae and myxomycetes, and many years ago the great German botanist Julius Sachs pointed out that each nucleus with its surrounding cytoplasm (what he called an ''energid'') was the functional equivalent of a cell. Furthermore, all these coenocytic organisms also have a cellular stage in their life cycle that has to do with the other main function for cells; they serve as units of reproduction.

The nucleus has a complete set of chromosomes, and if it is to reproduce into two, it must have a mechanism of equally dividing the genetic material into both daughter cells. As we know, this is achieved with extraordinary efficiency through the aegis of mitosis: the nucleus doubles or replicates its chromosomes, and one set goes to each of the two daughter cells, thereby preserving perfect genetic identity in the process. One notable exception to this rule is found in large ciliate protozoa that have a macronucleus that simply pinches in two with each division of the cell (Fig. 22). This process cannot continue indefinitely, even though the macronucleus has the basic genetic information duplicated many times, for errors creep in and the daughter cells become degenerate in various particulars. But rejuvenation is possible because there is also a micronucleus that carries a normal, double set of chromosomes and undergoes proper mitosis at each cell division. With a stable, preserved micronucleus it is possible to make a new macronucleus; this process occurs regularly after sexual fusion between partners, and it can even occur in an asexual reorganization. In both cases, the old macronucleus degenerates and a new one is formed from a perfect, intact micronucleus.

Ways of becoming multicellular

If, during the early course of evolution, unicellular organisms are to become larger, then the most obvious, and indeed the most successful method is to become multicellular. Such experiments in size increase, as with the ciliate protozoa discussed above, have quite severe limits because metabolism becomes ultimately ungovernable in a very large cell with a big, central nucleus. In any event, there are no organisms that have managed large size without becoming multicellular or multienergid. By becoming multicellular, an organism can preserve all the advantages offered by cells for efficient metabolism and proper gene distribution and at the same time become very large.

By far the most common method of achieving multicellularity is by the daughter cells growing and remaining stuck together so that a mass of cells is produced. The only real exceptions to this rule are a few small (but abundant) soil organisms that become large by the aggregation of

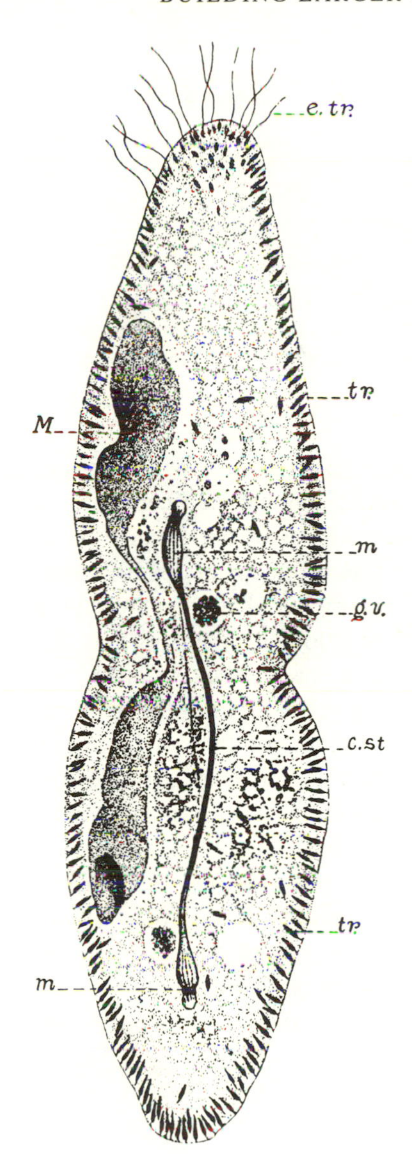

Fig. 22. A section through a dividing *Paramecium* showing the mitotic division of the micronucleus (m) and the pinching in two of the macronucleus (M). (tr., trichocysts; e. tr., extruded trichocysts; g.v., food vacuole). (From Calkins, *Biology of the Protozoa*, 1933.)

cells. In the cellular slime molds the amoebae feed and grow as separate individuals, and then, after starvation, come together by chemical attraction, or chemotaxis (Fig. 23). But this kind of development of multicellularity can produce minute organisms—minute, at least, when compared with dinosaurs and giant trees.

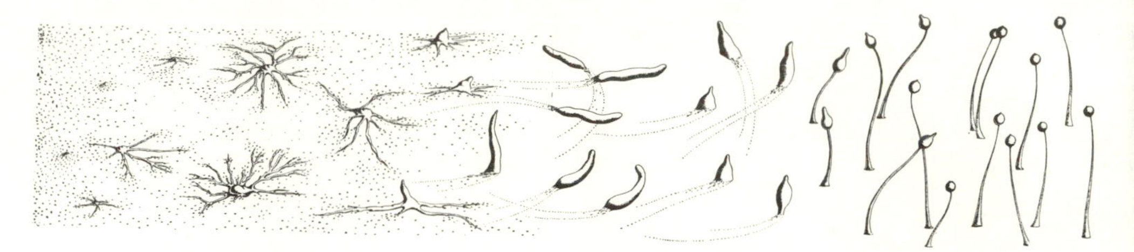

Fig. 23. The life cycle of a cellular slime mold (*Dictyostelium discoideum*) from the feeding stage (*left*), through aggregation, migration, and the final fruiting (*right*). (Drawing of Patricia Collins from Bonner 1969, Copyright © by *Scientific American*.)

If one looks at those organisms, primitive or advanced in structure, that live today, one must come to the conclusion that multicellularity is not something that arose once at some early time, but frequently, at many different times during the long course of the history of the earth. The main reason for suggesting multiple origins of multicellularity is because so many different kinds of small colonial organisms exist today, often producing similar or convergent shapes. For instance, one finds cell filaments of similar form among green algae and cyanobacteria, yet there is every reason to believe, because of the vast difference in their cellular structure, that the cell types themselves arose many millions of years apart (possibly 2 billion years) (Fig. 24). One could make similar arguments comparing cells with and without stiff secondary cell walls, for it is hard to imagine that sponges and algae have a common multicellular ancestor. But even among the algae, and no doubt among the other lower invertebrate groups, there is the possibility that there have been independent inventions of multicellularity because of the lack of separation of cell-division products. Finally, to further emphasize the

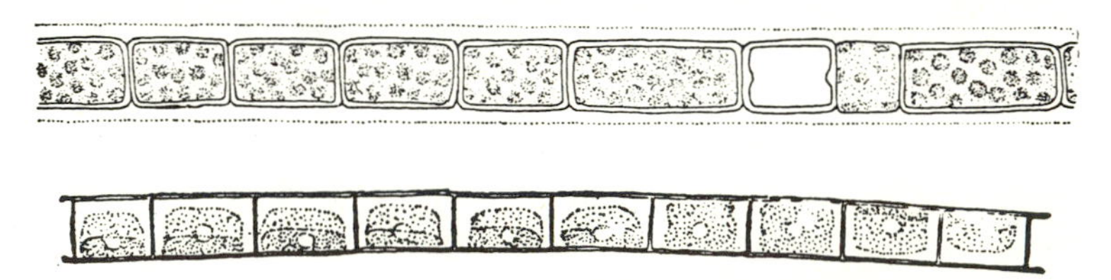

Fig. 24. A comparison of a filament of a prokaryote (the cyanobacterium, *Aulosira* × 725) *above* with that of a eukaryotic green alga (*Hormidium* × 975) *below*. (From Smith, 1950, Copyright © by McGraw-Hill.)

point of the likelihood of multiple origins, there is also good evidence for them among aggregation organisms. Aggregation is found in slime bacteria (myxobacteria), in two different types of amoebae (the acrasids and the dictyostelids), and in one species of ciliate protozoa. Again the evidence is that their origins are separate since the differences in these three cell structures are no doubt more ancient in evolutionary history than is their ability to aggregate.

SELECTIVE FORCES FOR MULTICELLULARITY

The fact that there has been a frequent repetition of the evolution of multicellularity forces us to ask the question, what could have been the selective pressures that might have brought this about?

There are three main reasons one could postulate for the origins of multicellularity. One is that a group of cells together can feed more effectively than separate, single cells. For instance, there is evidence that swarms of myxobacterial cells and the plasmodia of true slime molds (myxomycetes) can break down food by virtue of being large; they secrete quantities of extracellular digestive enzymes sufficient to break down large particles of food that are beyond the capabilities of single cells.

The second postulated reason is more effective dispersal. This is well illustrated in many small terrestrial organisms that produce small, stalked fruiting bodies containing a mass of spores at the apex. There are untold numbers of species of small fungi—molds and mildews—that do this in a modest way, and large mushrooms do the same thing on a far grander scale. Such small fruiting bodies are the rule for aggregation organisms; these also have arisen independently many times. The selective advantages of an effective method of dispersal is easy to see in principle, but it is harder to understand how these myriads of small spore-bearing structures really do promote greater reproductive success. There is something missing in our knowledge of the mechanisms of the process, although the evidence that it must be a potent selective force is overwhelming.

The third reason, while clearly of importance for large organisms, is more difficult to explain for small ones. It has to do with protection from predation and other external adversities; there is safety in numbers. Undoubtedly in some instances this played a part in selection for multicellularity, but it must be far less significant for these small cell colonies.

At least it is harder to suggest possible examples. For instance, one could imagine that for many aquatic forms, producing shells, as in the shelled amoebae, would be a more effective antipredator gambit than becoming multicellular. On the other hand, large, motile aquatic animals can swim faster than smaller ones, and this might be a good way to avoid predators. One can make other, similar hypotheses, but although all of them are possible, none of them are robust.

Besides selection, there is another entirely developmental reason for size increase, which has been suggested by numerous authors. They point out that if some step is added to a development it is less likely to affect the organism adversely than if a step is eliminated, and once this step is added it may be useful for other purposes; it may be the basis for further development. M. J. Katz (1986) calls this process "phylogenetic ratcheting." To give one of his examples, if more chromosomes, thus more DNA, is added to an organism by an accident, such a duplication is likely to be viable and remain stable, while a deletion of part of a chromosome will often be lethal. In other words, this is an obvious and well-known way in which DNA might be added and subsequently used for new purposes as the organism becomes larger and more complex. Katz also favors the notion that size increase in the evolution of the brain might have been accomplished by much the same system: an increase in the number of neurons (and therefore an increase in brain size) has led to an opportunity for increasing the complexity of the brain, while any decrease in nerve cells might be detrimental. It is easier to add than to subtract, and the additions can be the material for further developments. This would apply equally well to the origin of multicellularity. If cells come together by any means, either by chance aggregation or chance failure to separate after division, they are in a position, by having slipped into this new notch in the ratchet, to do something which capitalizes on the increase in size, and which in turn may end up as selectively advantageous.

WHY DEVELOPMENT?

It is easy to see that there may be numerous reasons why becoming multicellular might be advantageous, and here we ask why this requires development; why don't multicellular organisms just pinch in two, like single cells, and regenerate their lost size? Instead, almost all multicellular organisms begin each life cycle as a single cell, either a fertilized egg or zygote or an asexual spore. Since selection operates in the last analysis on genes, and since those genes are ultimately responsible for the activities of the cells within an animal or plant, it is essential to have a stage in which they either exist in two single copies, one from each parent (sexual organisms), or in at least one copy (asexual organisms). The mechanism of the handling of genetic information in chromosomes which undergo mitosis, meiosis, and fertilization is something that was invented and preserved by natural selection in the most ancient and primitive eukaryotes. Once this system was laid down, it could be put to use incorporating changes by selection, including size increase. If there has been selection for size increase in organisms, and there is the requirement for a single-cell stage dictated by the genetic system, the inevitable consequence is a life cycle which includes a period of size increase, a development.

Now, I would like to turn to some new aspects of the problem. Given the inevitability of development with increase in size, there are two consequences which will be examined in detail. One is that the larger the animal or plant, the greater the need for producing, during development, some system of support. The other is that as development takes longer and becomes more elaborate, it is increasingly bound by its own mechanics; each step in the building of the organism sets rules as to what steps can or cannot come next.

Development of support structures

If we begin with the structures of support, the first important point is that cell adhesion is an essential component for the multicellular condition; without it we would fall away and become rubble on the ground. L.E.R. Picken (1960) made the interesting observation that life cycles of multicellular organisms are an alternation of adhesive and nonadhesive states. Single-cell eggs and sperm that have been shed show no adhesion until the final moment of fertilization, and from then on the long, adhesive part of the life cycle takes over.

One can see the same kind of sequence in the asexual cycle of the aggregative cellular slime molds. There the amoebae feed as separate individuals, and when they come in contact they do not stay together, but slip past one another. However, as aggregation of the cells proceeds, they begin to secrete specific glycoproteins that are apparently directly involved in cell adhesion in different ways. There is even evidence for such adhesion molecules that do not appear until all the cells are together in a migrating cell mass or slug. Here and in animal embryos in general there has been considerable current interest in the chemical properties of these adhesive glycoproteins and the developmental timing of their synthesis.

Amoebae and animal cells in general have no cell wall, but merely a cell membrane which may contain at its surface the adhesive molecules. All plants, including many lower organisms such as algae, fungi, and prokaryotes (bacteria and the larger cyanobacteria), have cell walls. Material is secreted on the outside of the cell membrane that accumulates as a nonliving accretion. This gives the cell rigidity, and often immobility, converting amoeboid motion into protoplasmic streaming or cyclosis within the rigid cell. The cell wall may be gelatinous in consistency, as in some algae, or it may be stiff and composed of the polysaccharides chitin (in fungi) or cellulose, as in most other plants.

As one might expect, those algae whose cells are encased in jelly not only remain aquatic, but never achieve much size. A good example is the spherical *Volvox* in which the cells are separated from one another by jelly (although the cells remain connected to one another by protoplasmic strands). It is not surprising that with this relatively soft intercellular skeleton, *Volvox* only reaches a size of about 2 millimeters in diameter (Fig. 25).

Many algae which live attached to the ocean bottom in or just below the intertidal zone are subject to the continual action of waves. They would collapse instantly if their cells were held together by soft jelly; they need something much stronger. The large brown algae illustrate this point perfectly. Their cells are encased with a mixture of cellulose interwoven with a polysaccharide called alginic acid. This provides a combination of rigidity and suppleness that gives these large algae the consistency of a rubber garden hose, and as a result they can be very large (100 meters long) and live in turbulent waters (Fig. 26).

The red algae, which appear to be quite independent from the brown algae in their origin as multicellular organisms, have a mixture of cellulose and pectin in their cell walls. They do not achieve any great size, but some species have added a feature found in marine animals. These so-called coralline algae accumulate calcium carbonate, which gives them a very rigid, cementlike support (Fig. 27).

All these inventions by plants of supporting substances in the cell wall and their distribution around the cells are for aquatic organisms. In

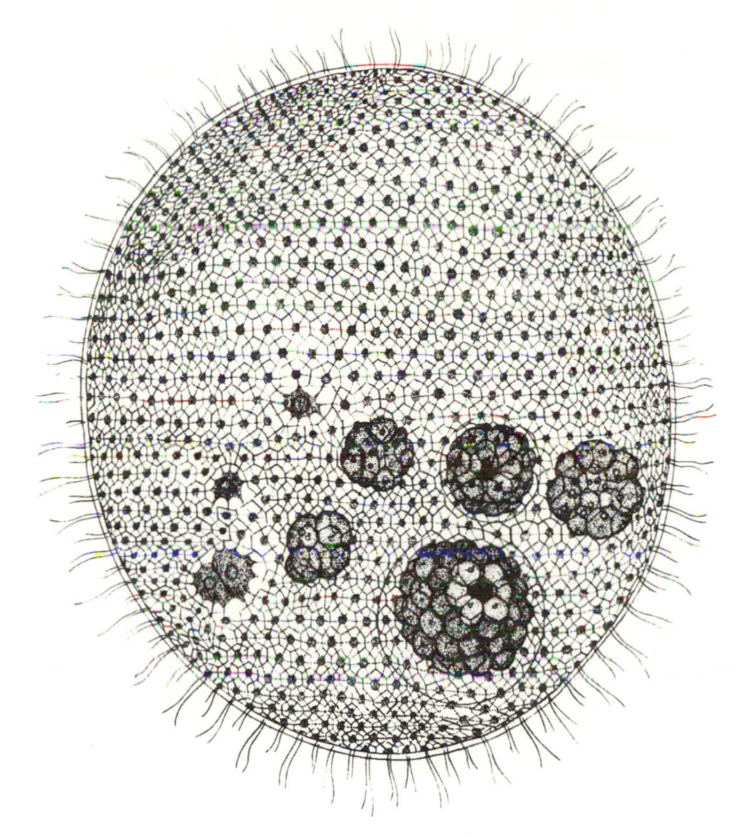

Fig. 25. A colony of *Volvox*, showing the successive stages in the asexual development of a daughter colony. (From Conklin.)

water, strength and the maintenance of multicellularity require only adhesion and sufficient support to resist wave action. Otherwise, the aquatic environment is relatively benign and undemanding of the support system. With the conquest of land, gravity became a force of major consequence, at least for large plants. However, the first terrestrial organisms were minute and well below the size where the effects of gravity could be felt. To examine the problems of these very small pioneers on land, let us use fungi and slime molds as examples.

The basic cell structure of fungi is the filament (hypha). There are species of terrestrial molds, such as the mucorales, that send their filamentous fruiting bodies directly up into the air (Fig. 28). This delicate,

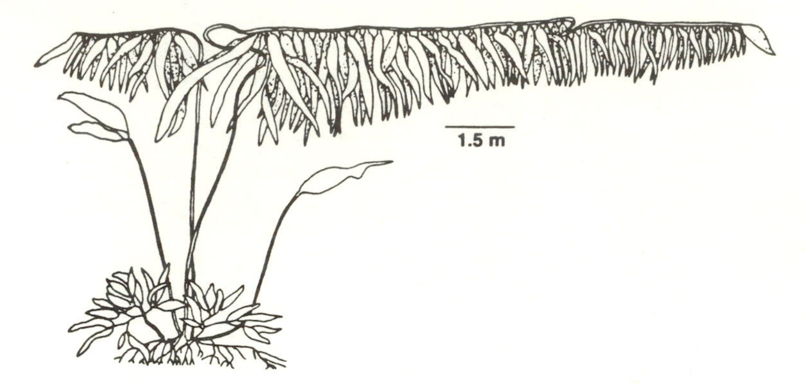

Fig. 26. The giant kelp, *Macrocystis*. (From Lee, *Phycology*, 1980, Copyright © by Cambridge University Press.)

aerial spore-bearing hypha is undoubtedly not affected by gravity; but it has to be able to break through the surface tension of a film of water as it begins it trajectory upward, and it has to be sufficiently rigid to remain upright despite air currents that might brush by it. This strength is achieved entirely because the cells are hollow cylinders. Such a shape is, as every engineer knows, an optimal one to resist buckling. The oriented units (micelles) of chitin in the walls give it the needed strength, so despite its small size it can remain stiff and erect even after being disturbed by small worms or insects.

In terrestrial molds one sees a trend towards size increase in fruiting bodies, and this has been achieved by the fungal filaments, or hyphae, growing beside one another in a fascicle so that one has a mass of filaments all oriented in the same direction (Fig. 29). Such a construction allows mushrooms to achieve remarkable size, some being half a foot high and often as much as 8 inches across the cap. Here the strength comes from the composite cylinders of pectin cell walls that lie parallel to one another. It is interesting that the tendency of the hyphae to grow adhering to one another is another example of Picken's principle of alternating adhesive and nonadhesive states. The feeding hyphae which sop up nutrients from the soil are separate and avoid one another, apparently by a repellent substance they themselves produce. But once the development of a mushroom begins, the hyphae suddenly reverse their habits and grow in parallel; the hyphae seem to be cemented together in the stalk of a mushroom.

The cellular slime molds provide a useful example because they have

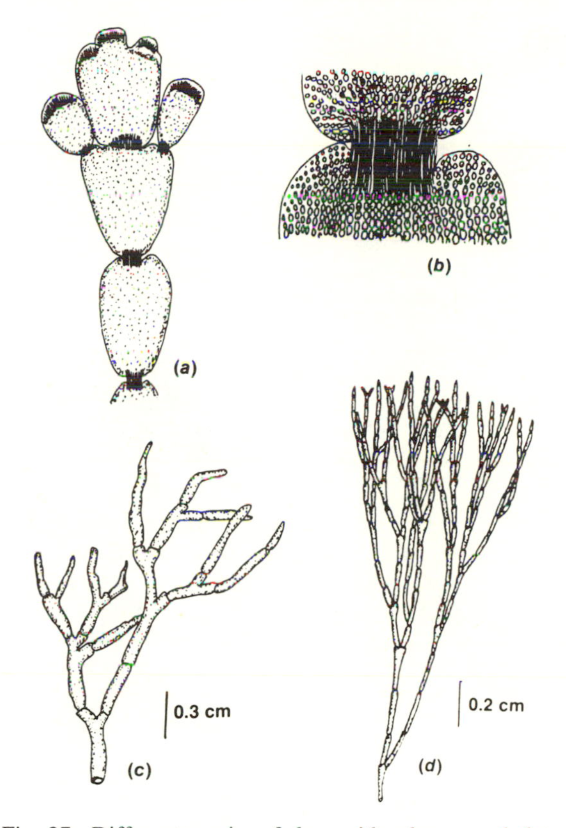

Fig. 27. Different species of algae with calcareous skeletons. (a) *Corallina*; (b) showing a noncalcified joint connecting two calcified segments of *Corallina*; (c) *Amphiroa*; (d) *Jania*. (From Lee, *Phycology*, 1980, Copyright © by Cambridge University Press.)

a combination of stages without cell walls, and then a stage where the cells are encased in cellulose. The amoebae have only flexible cell membranes, and when they come together by aggregation it would be quite impossible for them to rise into the air without some extra supporting material. There are two major ways in which they do this. In one genus called *Acytostelium*, the cells within the aggregated cell mass all secrete cellulose at their central (or axial) ends, with the result that a slender, noncellular cylinder of cellulose emerges out of the posterior end of the cell mass (Fig. 30). Since this cylinder is attached to the substratum by slime, and since the cells secrete it so that the tip of the

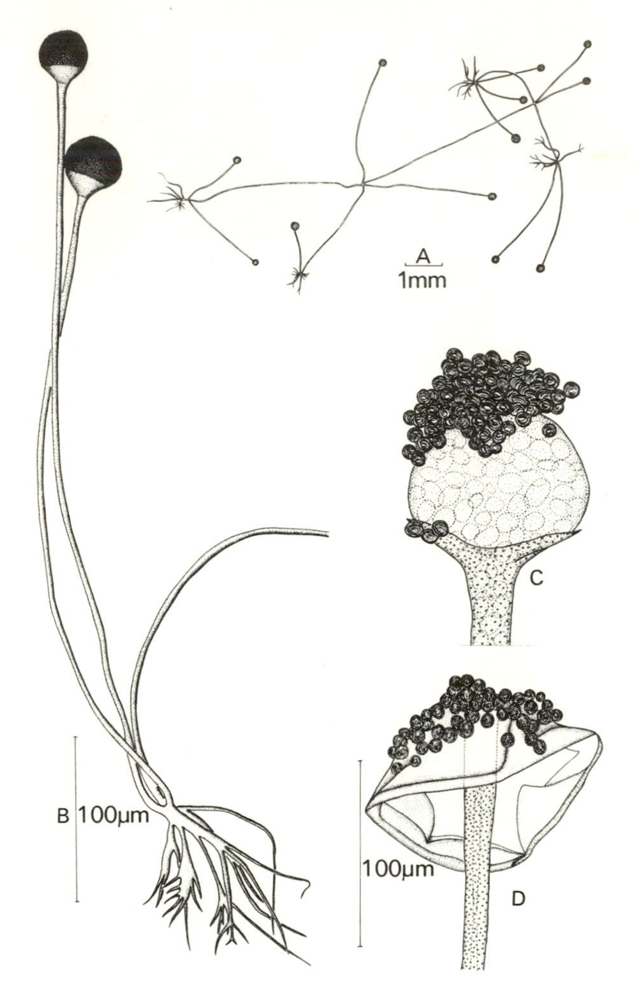

Fig. 28. The common mold, *Rhizopus*. *A*, the appearance of spreading filaments with spore-bearing bodies (sporangiophores); *B*, two sporangiophores with basal rhizoids; *C* and *D*, the spore-bearing tip of the sporangiophore. (From Webster, *Introduction to the Fungi*, 1980, Copyright © by Cambridge University Press.)

stalk is being continually added to, the cells rise up into the air and ultimately each cell turns into a spore.

In larger species of slime mold, such as members of the genus *Dictyostelium*, the tapering stalk cylinder encases cells. Here the cells secrete the cylinder as described before, but they rise up in a reverse foun-

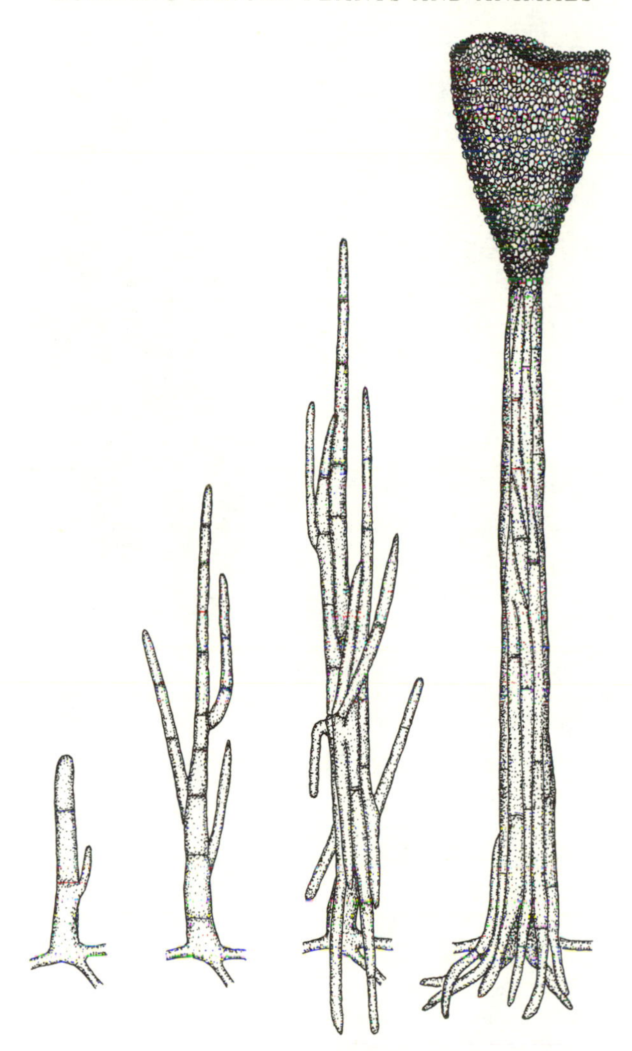

Fig. 29. Successive stages of development of a simple compound fruiting body of a fungus. (From Bonner 1952, Copyright © by Princeton University Press.)

tain to enter into the stalk tip that they have laid down. Once inside the tip, they start forming large vacuoles, and when the cells swell to a certain size they secrete a cell wall made of cellulose, and then die. The result is a stalk that has a diameter and length greater than that of *Acytostelium* as well as the added feature of greater support from the many

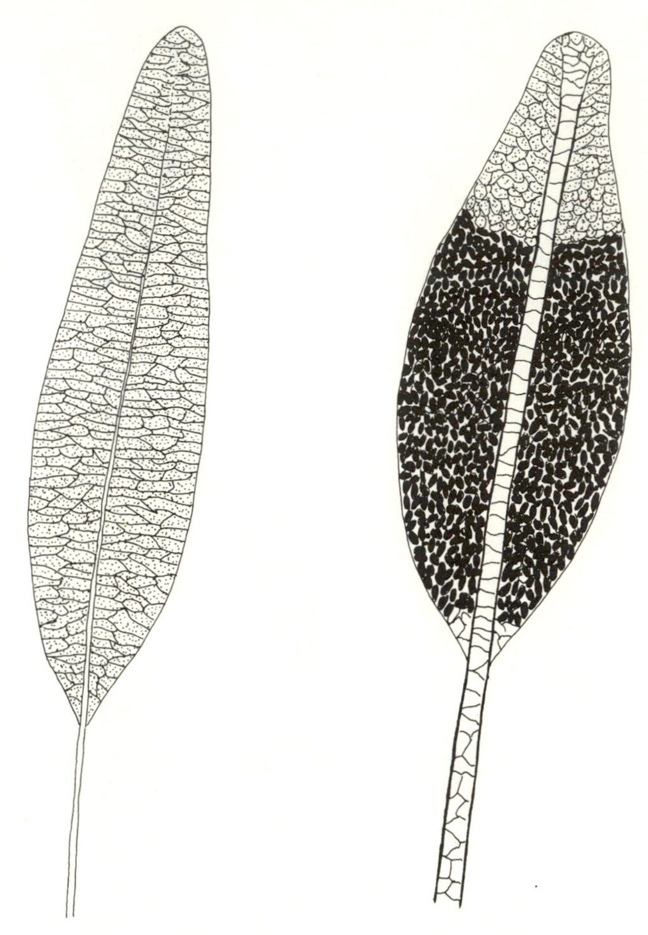

Fig. 30. Semi-diagrammatic drawings of saggital sections through the rising cell masses of two different cellular slime molds. In *Acytostelium (left)* all the cells secrete the noncellular stalk and then each cell turns into a spore. In *Dictyostelium (right)* the anterior cells form the stalk by first secreting the sheath and then entering at the tip by a ''reverse fountain'' movement. Once inside, the cells swell and die. The remaining posterior cells form spores. (*Dictyostelium* species are much larger than those of *Actyostelium*. In order to compare them, a small *Dictyostelium* is placed alongside a large *Actyostelium*.)

cross struts that result from the cell walls formed by the vacuolate stalks cells (Fig. 30). The *Dictyostelium* stalk is bigger and more elaborate, and one is tempted to think of it as having evolved from the more fragile *Acytostelium*; but such speculation, however tempting, is fraught with danger. It could as easily be true that the acellular stalk of *Acytostelium* is a specialization of the ancestral *Dictyostelium* condition, perhaps adapted to some particular environment. In any event, they both exist today (although *Dictyostelium* is more prevalent), and therefore both are assumed to be adaptive in their own way. From our point of view here, it is not the evolutionary sequence that is important, but rather the fact that there have been, during development, specialized constructions of supporting material invented by amoebae so that they can, as a group, stick up into the air.

The evolutionary line that has led to higher plants is presumed to have its origin in some of the green algae. Cellulose reigns supreme as the molecule for building cell walls, and the cylindrical filament evolved into the more rectangular plant cell, which became packed in a tissue. Such a cellular structure can be seen in liverworts and mosses, which presumably arose from green algae ancestors (Fig. 31). The bulk of these plants lie close to the ground, although they do produce either gametes, as in liverworts, or asexual spores, as in mosses, on stalks which rise up into the air (Fig. 31). This is another example of the importance of a raised dispenser of reproductive cells for dispersal. The internal cells of these stalks, with their cellulose cell walls, line up in parallel to give the kind of support already described in mushrooms.

As one looks from these small terrestrial plants to larger ones both in fossil forms and in their living representatives, one sees a progressive improvement in the strength of the upright portion of the plants. Ultimately, in higher plants one finds strands of vascular tissue in which the fibers have very thick cell walls that play a key role in giving strength. In trees, these tracts of fibers with their accompanying system of transport tubes turn into wood; and as this happens lignin, another cellulose-like polysaccharide which provides added rigidity, is deposited in the secondary walls of the cells. The cells of wood, after all this deposition of strengthening molecules, die in much the same way that the stalk cells of *Dictyostelium* die. There are many variations in the details of the structure of different vascular plants, but the principal is the same for all; the rigidity comes from the cell walls, and these rigid cells are slowly accumulated during a prolonged development to form wood. This is what permits giant sequoias and eucalyptus trees to gain heights greater than a hundred meters.

Considerable interest has been shown in the supporting structures of a number of aquatic animals, especially by S. A. Wainright and his associates (1976) and others. The subject is too large to go into in any detail; here I will choose a variety of different examples of ways in

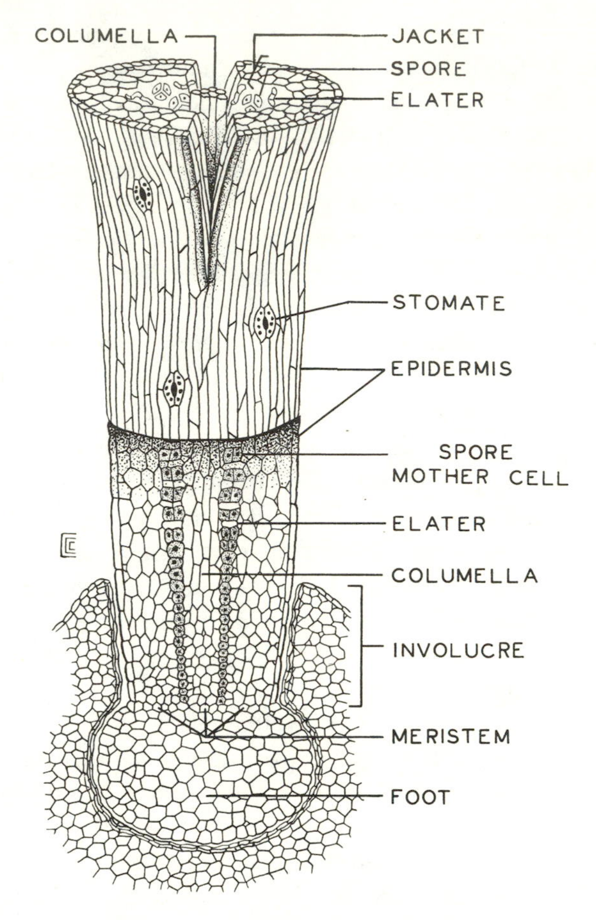

Fig. 31. Drawing showing the cellular structure of a primitive bryophyte (*Anthoceros*, a liverwort). (From Cronquist, *Introductory Botany*, 1971, Copyright © by Harper & Row.)

which the animals have provided themselves with supporting mechanisms, beginning with lowly sponges and cnidarians (coelenterates), and following up with and going beyond the conquest of land, as was done for plants.

Sponges appear to be very loosely held together, for the cells can be easily separated by any kind of rough treatment; yet sponges achieve considerable size, and some live in places where currents put a constant stress on them. They achieve their support by building, during their development, extracellular spicules, and these can be made of a flexible polysaccharide called spongin (as in a bath sponge), calcium carbonate (the calcareous sponges), or silica (the glass sponges and desmosponges). The spicules are secreted by specialized spicule-forming amoebae that travel up and down the spicule as though they were painting on the skeletal material (Fig. 32). The question of how the spicule achieves a certain shape is a fascinating subject summarized in all its splendor by D'Arcy Thompson in his famous book *On Growth and Form* (1942) and studied by others since then. It is, however, peripheral to the central point made here, so I will just say that the external shape of the spicule is determined by the cells, and not by crystallographic forces. Whatever the shape of the spicules or the material from which they are made, as development proceeds they become interlaced between the cells and form a firm fabric upon which the many cells can

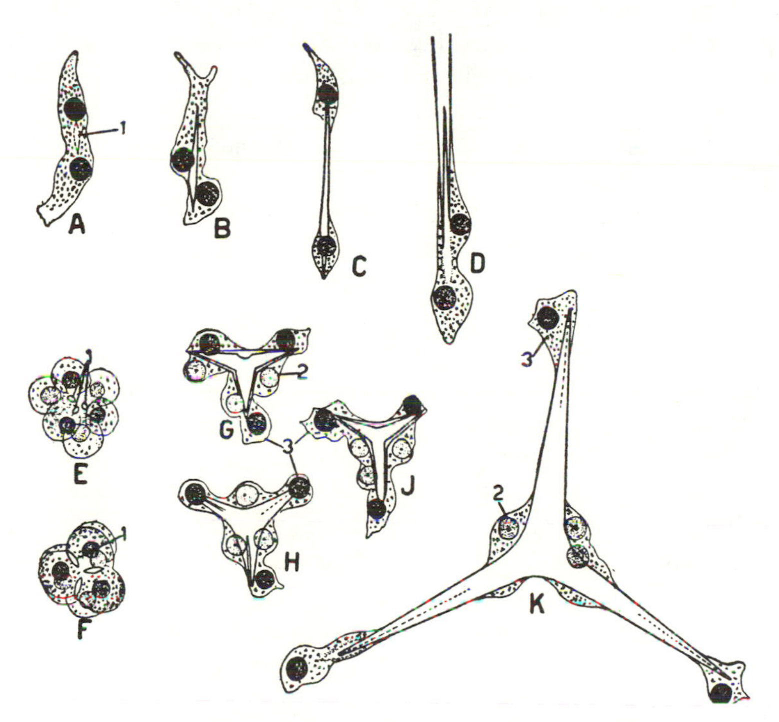

Fig. 32. The formation of spicules in sponges involving the secretion of calcium carbonate by special spicule-forming amoebocytes. (1, the beginning of a spicule; 2, amoebocytes that deposit near the axis; and 3, those which deposit at the spicule apices.) (From Hyman 1940, Copyright © by McGraw-Hill.)

attach themselves and provide the necessary support. In this way sponges take on all sorts of interesting shapes, and some of them will be quite large (Fig. 33).

Among cnidarians we find a variety of different ways support has been achieved. Among the jellyfish, the jelly itself has become the support. It is not living tissue, but a great mass of inert gel that lies between the thin external and internal covering layers of cells. Unlike algae, some jellyfish grow to a remarkable size, the largest being of the genus *Cyanea*, which may have a bell over 2 meters in diameter. They do not stand still and resist the currents like sponges, but are wafted along. Their own swimming movements play a more important role in going up and down in the ocean, possibly to find the zones rich in food.

Among the smaller, more primitive cnidarians, the stalked hydroids, one finds a number of species that secrete during development either a thin plastic-like sheath (made of chitin) around the stalk or a thick calcareous deposit, as in corals (Fig. 34). Some of the largest hydroids have such supporting exoskeletons, which also provide protection from predators. It is difficult to know in any particular instance which role is more significant; it is a case of two selective advantages provided by the same structural device.

In the more advanced phyla of invertebrates one finds a number of new inventions which are built during development to provide support. In many of the soft-bodied organisms, such as the different phyla of worms, it is the thickened sheets of muscle that not only permit active swimming, but provide the support. The muscles attach to the membranes, and these in combination give a form of support. However, in order for muscles to be most effective, they should be attached to something rigid. The first big advance in this direction is the appearance of an exoskeleton during development. Exoskeletons have arisen independently in a large number of invertebrate phyla (besides the cnidarians already mentioned), such as molluscs, bryozoans, brachiopods, echinoderms, and arthropods.

In the arthropods the skeleton provides sufficient support for very large size, as in king crabs, as well as for terrestrial and even aerial forms. Since the skeleton, which is made either of chitin or chitin encrusted with calcium carbonate (lobsters and crabs), is secreted during development by cells in the covering layer of the animal, it must be remade at each molt. Therefore, from a developmental point of view, an arthropod's exoskeleton is expensive, but it does provide both an

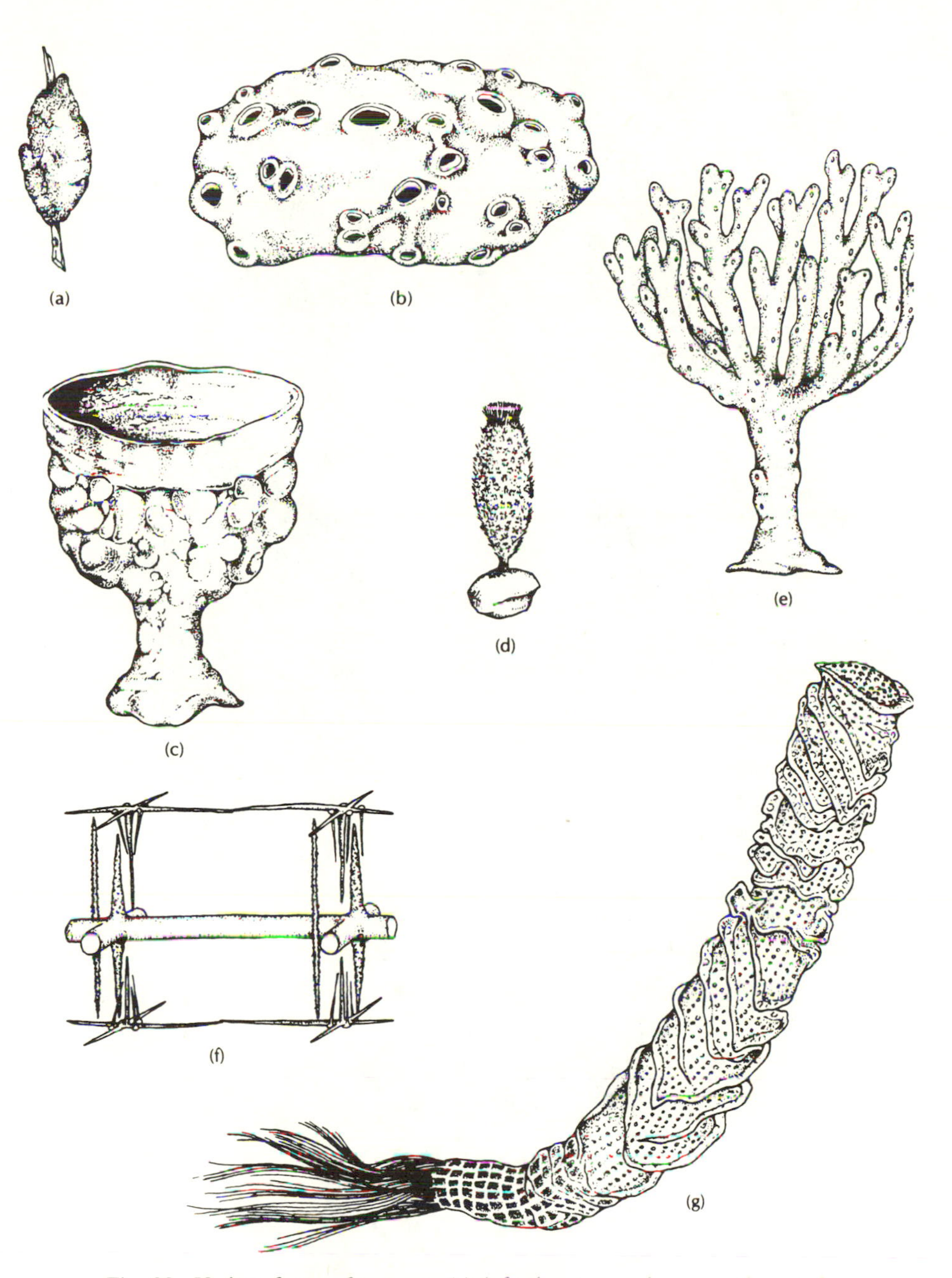

Fig. 33. Various forms of sponges. (a) A freshwater species encrusting a twig. (b) An encrusting marine sponge. (c) Neptune's Goblet sponge. (d) A syconoid sponge, *Sycon* sp. (e) A fingered sponge, *Microciona* sp. (g) *Euplectella* sp., Venus's Flower Basket. This is a member of the Hexactinellida. The spicules are six-rayed and composed of silica, as shown in (f). (From Pechenik, *Biology of the Invertebrates*, 1985, by permission, Wadsworth.)

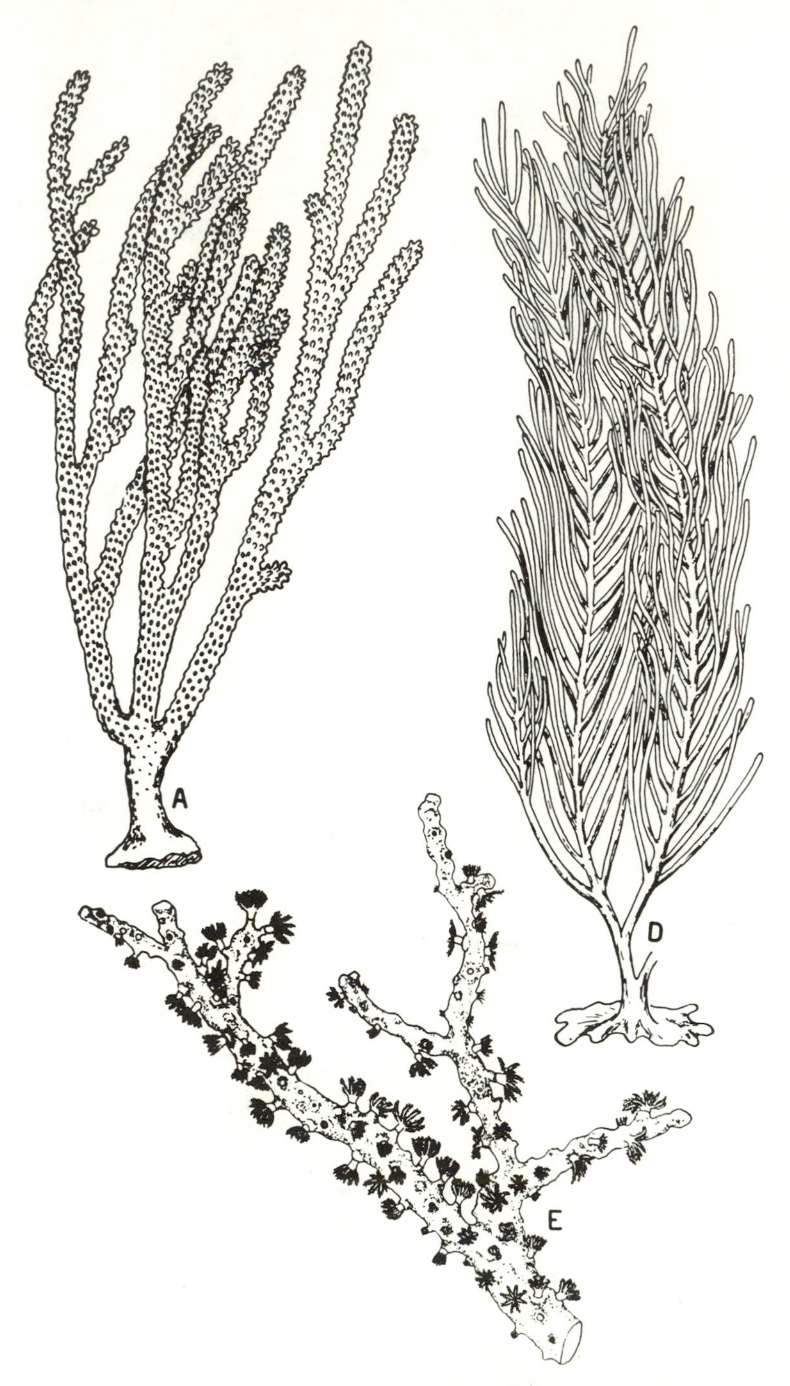

Fig. 34. A variety of shapes among corals. *A*, a gorgonian coral; *D*, a sea-whip gorgonian; *E*, Mediterranean red coral. (From Hyman 1940, Copyright © by McGraw-Hill.)

excellent frame for support and an armor for protection (except during the vulnerable period of molting).

An even more successful form of support is provided by an endoskeleton, a structure that appears sporadically among different invertebrate groups, such as the pen or cuttle found in some cephalopod molluscs like the squid. But the great leap forward came with the development of a notochord in protochordates. Here was a structure that in its earliest appearance in evolution probably held its rigidity by turgor pressure, and only later, we presume, did turgor become replaced by the accumulation of cartilage and bone. From the developmental point of view, a notochord is laid down very early in the embryo, after the formation of the neural tube has begun. As evolution proceeded, processes of bone or cartilage began to shoot off from the axial chord, and ultimately these processes became more elaborate, forming the bones of the hands and feet of amphibians and of all higher groups of vertebrates.

It is interesting that this endoskeleton also came to surround the enlarging brain in the form of a skull. Here the endoskeleton has become what amounts to an exoskeleton, although fortunately vertebrates do not have to molt their skull as they grow, for growth can occur at the edge of the bone plates of the skull. (This is also true for the exoskeleton of echinoderms, such as the test of a sea urchin.) This vertebrate "exoskeleton" certainly serves the purpose of protecting the vital brain tissue, but it also helps to support that tissue, which is especially fragile and delicate.

The notochord and its ossification, and the formation of peripheral elements in the form of ribs and fins, were all achieved by aquatic organisms. In water the support provides more effective propulsion because the backbone and the other bones allow points of attachment to the muscles so they can produce effective swimming motions. During the early evolution of the vertebrates, this was a fortunate bit of body design, because with the conquest of land the endoskeleton was able to adapt beautifully to the problem of supporting weight against the pull of gravity. No doubt the first terrestrial animals were small and the force of gravity was negligible; but with size increase gravity became important, and the body plan, set down in early development, was ready.

The changes that have taken place in the architecture of the skeleton for all vertebrates, living ones as well as those fossils in which the skeletal remains are sufficient for a clear reconstruction, can be related primarily to the posture of the beast and to its manner of locomotion. In this case, the structure of the skeleton is an essential feature of the phenotype, and presumably the major changes in that phenotype have come about by the standard process of natural selection. Yet those gene-controlled determinations of structure are always bound by strict limits set down by the basic principles of engineering. Let me now illustrate this point in an important way, a way which applies to both large terrestrial animals and plants.

Proportions and Size

In terrestrial vertebrates, we know that with larger species, the bones become disproportionately thick. The relation is well established and can be expressed in very precise terms. Furthermore, the same relation holds for large trees; the effects of gravity are the same for large terrestrial animals and for plants.

The simplest way of expressing the exact empirical relation between size and proportions is to say that the diameter of the bone or tree varies as the length to the 3/2's power:

$$d \propto l^{3/2}.$$

If there were no disproportionate increase in diameter and the small and large bone (or tree trunk) were geometrically similar, then the diameter would be directly proportioned to the length:

$$d \propto l.$$

If one measures bones or trees of different sizes, one can see that the $d \propto l^{3/2}$ relation holds (Fig. 35).

The most ingenious insight into the possible physical cause of this relation comes from T. A. McMahon (1973). He points out that this is what one would expect if the structures of differing size were to keep the same elastic properties. This principle of elastic similarity is one of pure engineering; it would apply to inanimate objects with equal force. It is also possible to show that very minute organisms do not show elastic similarity but geometric similarity, so elastic similarity is indeed an accommodation to the force of gravity (McMahon and Bonner, 1983). However, although the proportions of many structures from bones to trees of differing size follow the 3/2 relation, there are many exceptions to the rule. This has led some to question the generality of the elastic similarity model.

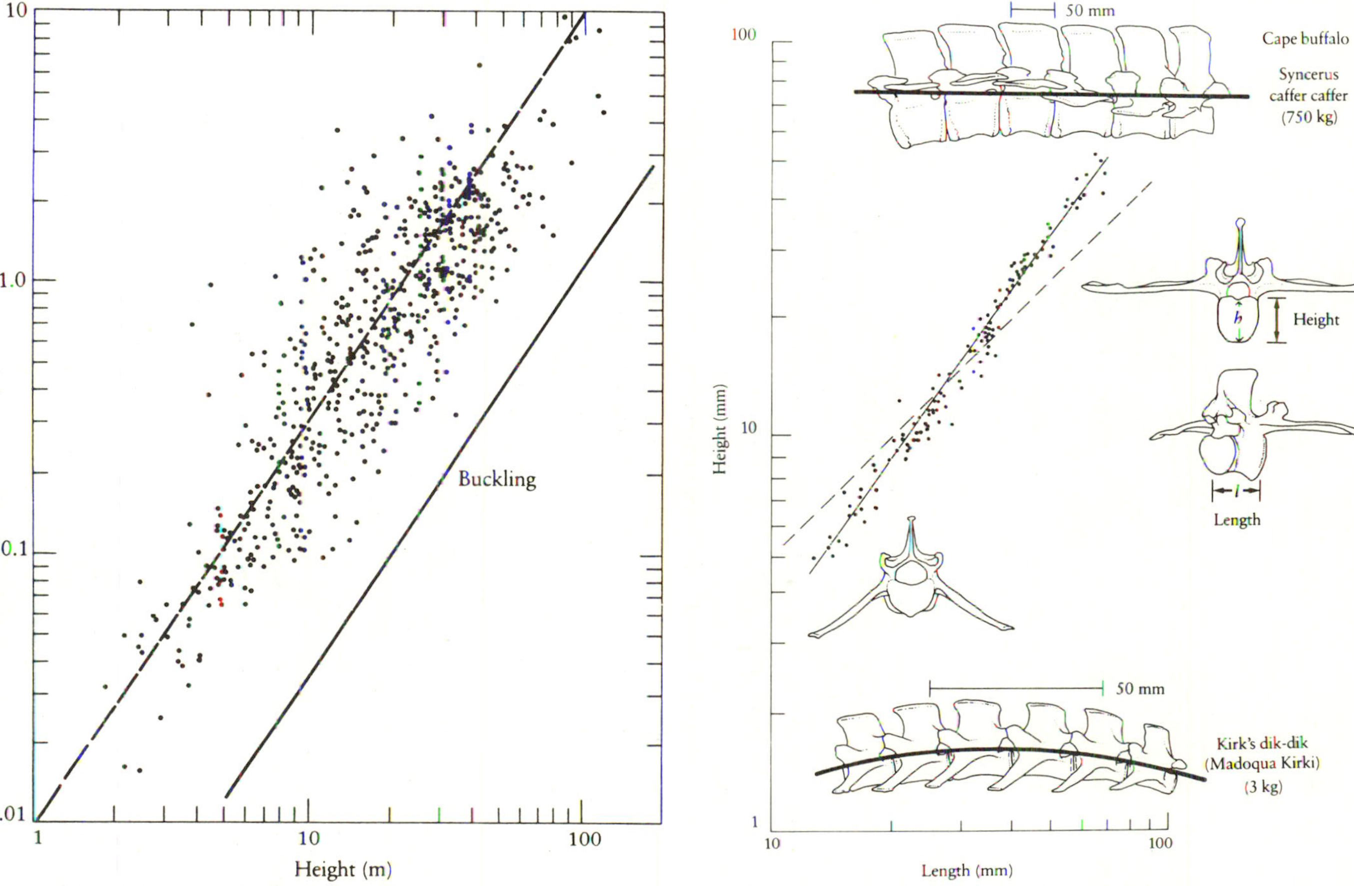

Fig. 35. *Left*: Diameter of the base of different trees plotted against overall height on a log-log graph. These are 576 record specimen trees, representing what is believed to be the tallest and broadest of each species found in the United States. *Right*: Comparison of the centrum height and the centrum length for the third lumbar vertebra in African bovids, shown in a log-log plot. The centrum height (the diameter) is proportional to the centrum length raised to the power 1.47. A line of slope 1 (isometry) is drawn for comparison. (From McMahon and Bonner 1983, Copyright © by *Scientific American Books*.)

We must now ask the question of how the 3/2 relation arose during the course of evolution of large animals and plants, which leads directly to the question of how this relation is established during development. First, let us look at the case for animals. In mammals, the offspring at birth is fully formed and the long bones such as the femur are clearly well developed. The diameter of these bones stops increasing long before the length of the bone ceases to grow. Therefore, late in their growth these bones become relatively thinner with size increase and do not continuously follow the principle of elastic similarity. The only possible explanation of this fact is to assume that the plan for the adult length-diameter relation is preset and begins by establishing the correct diameter early, and then much later at the end of the period of youthful growth the long bones cease to extend at their ends. This means that at two moments in development there is specific gene action regulating growth, so that ultimately the appropriate proportions are achieved for the weight of the adult animal. Selection on the adult phenotype has produced the right gene signals appearing at the right time in earlier development.

The way in which the $d \propto l^{3/2}$ proportions are established during the development of plants is very different. It is easy to measure a tree as it grows, as was done by R. Kiltie and H. S. Horn for a number of species. For instance, American beech trees become disproportionately thicker as they grow, and they remain well below the buckling limit (Fig. 36). This kind of development of proportions is unlike animal long bones in that the beech's proportions are corrected at all times during development, while in mammals the proportions are achieved only at the end of development. This difference probably has something to do with the fact that trees are continuously subject to the forces of gravity during their development; they always stand free and unsupported. On the other hand, in mammals the fetus *in utero* is suspended in an aquatic environment, and gravity comes into play only at birth. However, the long bones of newborn mammals are greater in their thickness-to-length ratio than those of adults, so that they have more than enough support as they grow.

In mammals we postulated two moments of gene action for growth cessation of width and length, respectively. In trees there are various possible ways the genes could regulate the change in proportions with increasing size during growth. The simplest would be to set differentially the rates of the two growth zones, or meristems, so that the appro-

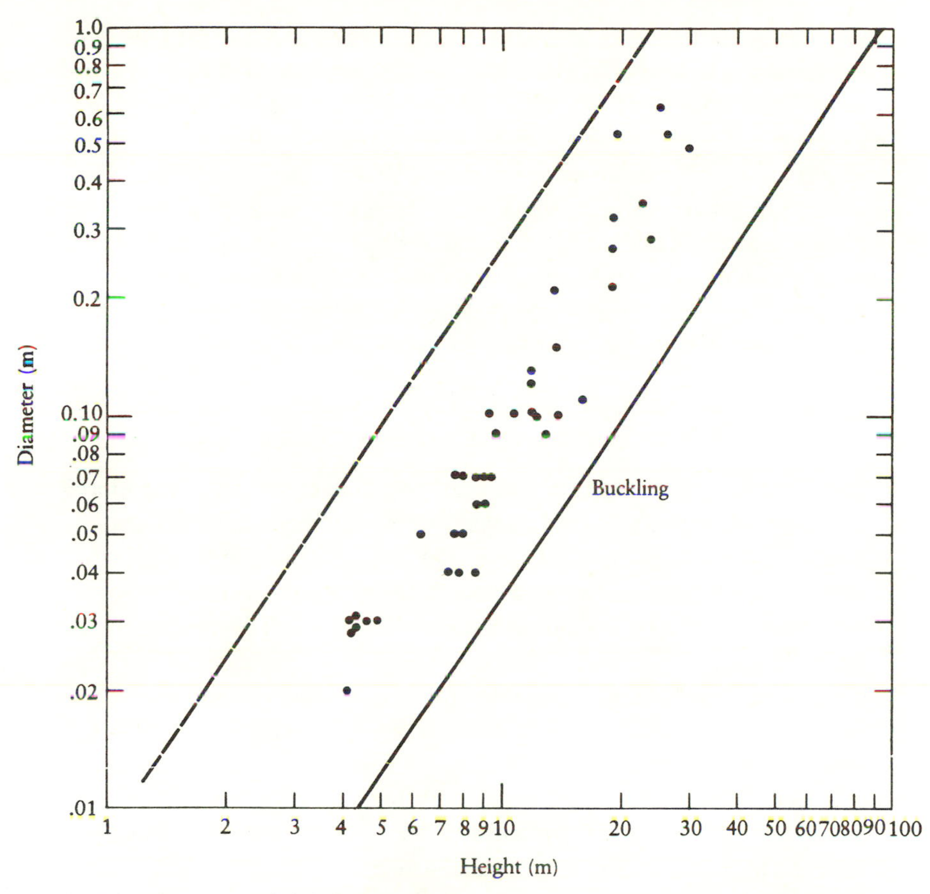

Fig. 36. Diameter vs. height in a series of American beech trees (*Fagus grandifolia*) of different ages, showing that diameter increases faster than height with increasing age. Both the broken line and the solid line are the same ones plotted in figure 35. (From McMahon and Bonner 1983, Copyright © by *Scientific American Books*.)

priate ratio exists at all times. Elongation at the tip is achieved by an apical meristem, and the lateral growth by a circumferential meristem called the cambium. The cambial growth could be so set that it would slightly outdistance the growth of the apical meristem, thereby producing a relatively greater increase in girth than length as the tree becomes larger.

As is evident, I am making no attempt here to identify the actual genes involved, or even to suggest how to look for them. All I want to do is to give a plausible, hypothetical scheme which marries the forces of natural selection with the 3/2 rule. First, we must make the reasonable assumption that bones and tree trunks are best off as slender as possible. This is required for agility in animals and for strength without excess weight in plants. But if they are too slender they will fail by buckling. Therefore we assume that different-sized bones and tree trunks are limited in their slenderness by elastic properties, and that natural selection calls for those organisms that are neither too thick nor too thin, that is, those that follow the 3/2 rule. Such selection means that genes lay down the appropriate rules for growth so that not only the adult is well proportioned, but also the young when it needs the support.

Developmental steps and size

Increased size of an adult produces not only longer periods of development, but also involves more steps, more processes, and obviously a much greater accumulation of mass. This means that a simple algal colony of a few hundred cells has quite a different set of construction problems than that of a large plant or animal. Here we will first concentrate on those larger forms and try to understand how long construction sequences are mechanically possible and what factors play a significant role in determining variations in the sequences. It is very much the same problem as constructing a large building: one cannot put in the windows or construct the roof before one has first laid down the foundation and put up the entire frame of the house. What are the rules of construction imposed by size increase during the course of evolution?

The first point of major importance is the nature of the building materials: the stiff cells encased in the walls of plants make a very different kind of building material from the soft cells of animals. Let me discuss plants first.

I have already pointed out that fungi can produce large mushrooms by having individual hyphae grow alongside one another in a fascicle. This construction will produce both the stalk and a cap with gills. Here all growth occurs at the hyphal tip, at least to form the original distribution of cells in the minute primordial button. Later, when the button turns into a large mushroom, those original small cells have become inflated with protoplasm that lies in the vegetative mass of filaments in the soil below. So the second phase is not growth in the sense of synthesis of new protoplasm, but an inflation of the cells which were earlier laid down in the appropriate pattern; and the mass transfer of protoplasm is responsible for the final, inflated size of the mature mushroom. This is one way that rigid cells can produce a large structure, but it clearly has its limitations, which may account for the fact that mushrooms achieve only a fraction of the height of even the smallest trees.

In trees the degree of rigidity of the cells is understandably much greater than in fungi, and there is no question of inflating cells, although to a minor extent this can happen in growing cells in an apical meristem. In every elementary biology laboratory we have the opportunity to admire an onion root tip which shows, near the very tip, cells that appear quite square and are often in the act of dividing (Fig. 37). This is the zone of multiplication. Above it the cells can clearly be seen getting longer, although their diameters remain fixed, and this is the zone of cell elongation where the cells swell, in this case due to osmotic changes accompanied by the formation of a large vacuole. Perhaps this is related way back in evolutionary history to the cell elongation of fungi.

The important point is that as a plant becomes large, and as its cells become rigid by further secretion of cellulose after elongation, the way in which a plant can grow becomes severely limited. This is especially obvious in the case of trees where the bulk of the tissue is dead wood, extremely hard and permanent; there is no way it can mold itself into different shapes. To get around this difficulty imposed by cell rigidity, the plant grows only in peripheral regions and partly by accumulation of cells from these meristems. For instance, expansion of the girth of the tree is the result of a thin cambium that surrounds the wood, and only here can cells divide and contribute to girth. All daughter cells of the dividing cambium that are inside turn into wood (xylem), and those on the outside turn into a thin layer of bast (phloem). The other meristems of large plants are the apical meristems, of which the onion root tip is a classical example. Here the tips of the shoot and the tips of the root become extended, while the cambium makes the tree thicker. It is the activity of these two meristems that mold the shape of the tree. This is a grossly oversimplified picture, but it illustrates the main point: plants can increase in size, and in fact become enormous by virtue of the invention of meristematic growth. But such growth also severely limits the kind of shape changes that are possible as growth proceeds. In particular, the shape at a given stage must be a relatively inflexible core for any later shape.

In the case of animals, we have an entirely different set of construction problems because of the lack of a cell wall. If we compare higher

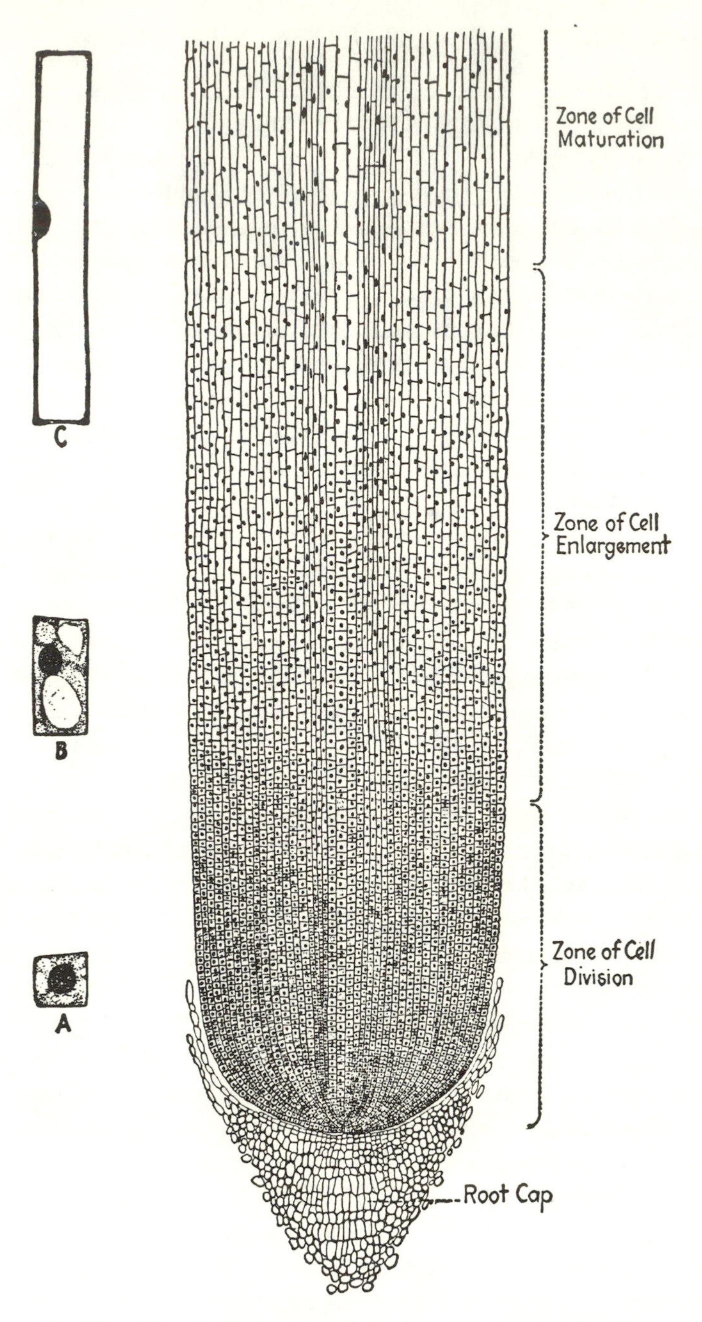

Fig. 37. A longitudinal section through a growing root tip. Note that the cells divide in the zone nearest the tip, and above that the cells elongate (zone of enlargement) and the two processes are thus separated in time and space. (From Eames and McDaniel, *Introduction to Plant Anatomy,* 1925, Copyright © by McGraw-Hill.)

plants and animals they also differ in that the plant, as a very young seedling, begins a pattern of development that lasts continuously until it is felled as a large tree. In large animals there is a long period where the embryo is protected *in utero*, or in a large shelled egg, or in an egg sac, and in this state of suspended protection it can undergo developmental changes that might be difficult or impossible if it were simultaneously trying to swim or feed.

However, we must remember that the ancestors of such large, advanced organisms probably had swimming larvae, and therefore the early developmental stages of the first multicellular organisms may have had no such period of protection. Let us begin with an examination of developmental construction problems of *Volvox*, sponges, and cnidarians before we examine vertebrates.

The case of *Volvox* is instructive because it shows how cell polarity can impose a restriction on development. If an asexual colony arises inside a mother colony, it is the result of numerous cleavages producing a sphere of cells that bulges inward (Fig. 38). The cells are polar in that they have a flagellar end, which in the adult protrudes towards the outside of the colony and is responsible for the locomotion of the entire colony. As a result of the inward bulging, in the daughter colony the flagellar end of all its cells are pointing into its interior cavity, making locomotion impossible. The problem is solved by the juvenile colony turning itself inside out, very much like turning a sock inside out (Fig. 38). We need not concern ourselves here with the mechanism of this turning, but just point out that there is good evidence that this so-called inversion is achieved by shape changes of the individual cells, which occur together in a moving, coordinated band. When the young colony is eventually liberated from inside the disintegrating mother colony, it is free to swim away.

In calcareous sponges, O. Duboseq and O. Tuzet (1935) showed a similar inversion in early development which so closely resembles *Volvox* that L. H. Hyman (1942) suggested (perhaps somewhat rashly) that *Volvox* and sponges are related. The result is the same in both cases: the production of a free-swimming flagellated blastula or hollow ball of cells that is the larva in sponges. This motile larva is a means of dispersal for the sponge, and by settling elsewhere it can form a new sponge. But in so doing, a second remarkable event occurs: the hollow blastula now inverts again, so that as it settles the flagellated cells return to the inside (Fig. 39). They now become the internal collar cells which drive

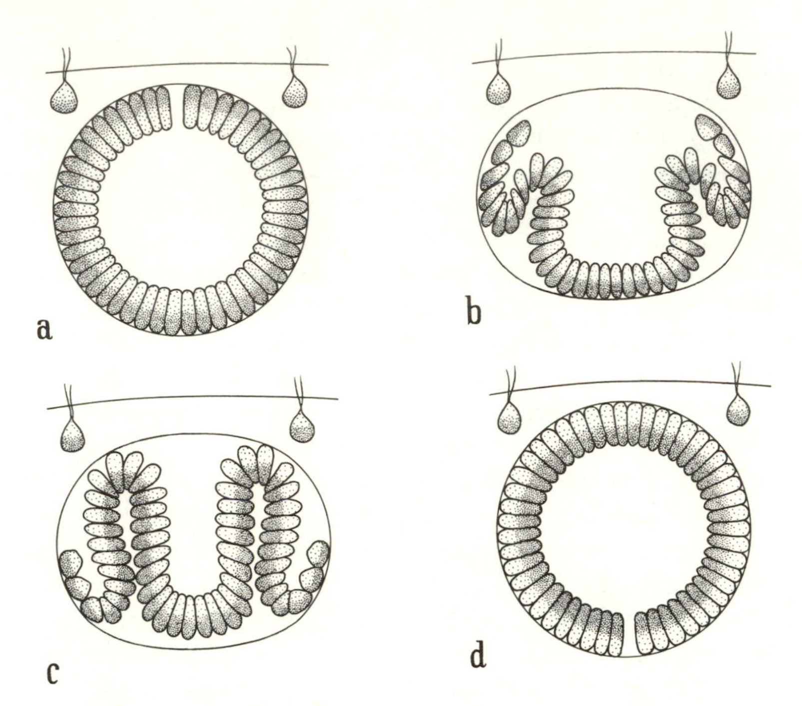

Fig. 38. A diagrammatic view of inversion in *Volvox*. The flagella sprout from the light ends of the cells. Note that at first the flagella ends of the cells are pointing inward, but after inversion they point outward. (From Grell, *Protozoology*, Copyright © by Springer-Verlag.)

the currents in the canal system within sponges, for a sponge obtains its food from particles in the water that their flagellated collar cells waft through its body.

The moral here is that development is under two influences: one is determined by rigid, given cell properties, and the other by the need to function effectively at all stages of development. In the case of *Volvox* and calcareous sponges, the cell polarity is fixed and cannot be reversed within the cell. But in order to swim, these organisms need their flagella outside, and both *Volvox* and calcareous sponges do this by the inversion of the ball of cells. Later, sponges need their flagella on the inside in order to permit feeding, and this they do by a further turning inside out. Function and the inherent, rigid cell properties require this complex developmental solution.

The next step in evolution was to produce animals of greater bulk,

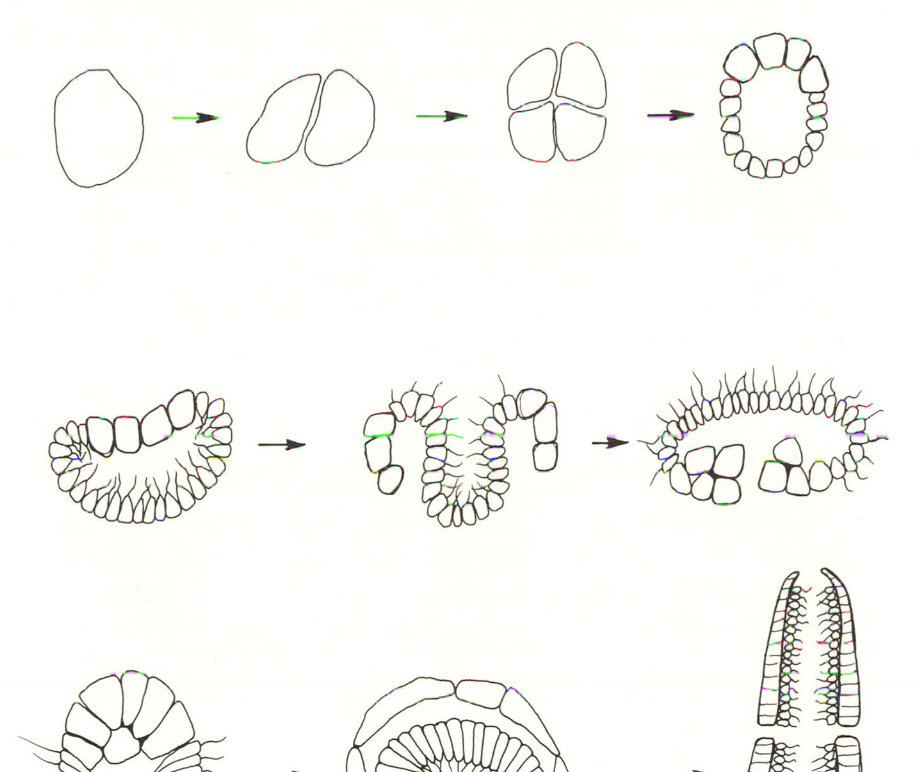

Fig. 39. A diagrammatic representation of the development of a calcareous sponge. After cleavage (*top row*) inversion occurs (*middle row*) to form a swimming larva with the flagella on the outside. Later, once the larva settles (*bottom row*), gastrulation occurs and the flagellated cells point inward to begin the circulation needed for feeding.

and much is made in animal embryology of the laying down of a second and a third layer of cells so that a body has an outside covering of cells (ectoderm), an inside layer (endoderm), and a middle one (mesoderm). The process of becoming thick with cells and cell layers begins with gastrulation, which can occur in many ways: it can be by the classic invagination, where a *Volvox*-like gastrula pushes in at one side to form a double layer of cells, or it can be by the wandering in of cells from the outer blastula layer to fill the blastula cavity. In both cases, it is the first step towards a body plan that can become large; it is the beginning of accumulation of cell mass.

In the hydroids, which are a major group of cnidarians, the fertilized egg cleaves into an elliptical blastula of cells, and some of these cells wander inwards to fill the central cavity (Fig. 40). The outer cells sprout flagella, and this planula larva disperses by swimming. It ultimately settles on the ocean floor and begins to develop a mouth at one end, while the internal mass of cells hollows out to form a central gut which becomes lined with cells. It is now a functional organism with a mouth which can capture and engulf passing food. The middle layer (mesoglea) is very thin and inconspicuous—less evident than in sponges. Yet in the related jellyfish, the massive, noncellular jelly is the middle layer.

If one proceeds up the evolutionary scale in size, there is the innovation of an anus, allowing food to go down the gut in a unidirectional fashion and be digested with greater efficiency. During the development of an animal the appearance of the complete gut is an important event, because before that moment the embryo cannot feed and take in energy. The embryo must, up to that point, rely entirely on the nutritious yolk deposits laid down by its mother in the egg.

In the vertebrates the problem of getting energy to the early embryo has been solved in a number of different ways. In fish, reptiles, and

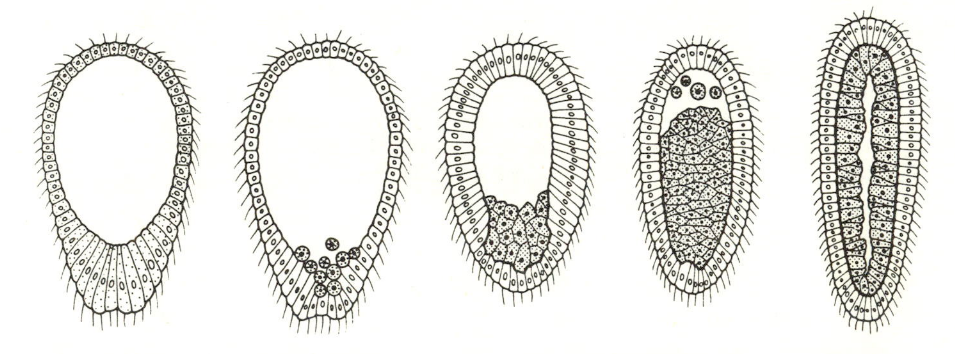

Fig. 40. In the hydroid *Aequorea* gastrulation occurs by the movement of cells inward at one end of the embryo (polar ingression), in this way producing a double-layer ciliated larva (right-hand drawing). (From Mergner in *Experimental Embryology of Marine and Fresh Water Invertebrates*, G. Reverberi, ed., 1971, Copyright © by Elsevier.)

birds there is a relatively large amount of yolk laid down, sufficient to take the embryo to a stage where it can feed itself. An extreme case would be that of a duck where the duckling emerges and can immediately swim about and feed.

Thus far I have described two important constraints that have affected the development of animals during the course of their evolution. One is the need for some embryos of small animals to move for dispersal; the other is the need to feed once the yolk reserves have been exhausted. These are special, primitive cases of a more general principle: each stage of developmental size increase has to be viable and adaptive. Besides managing ways to move or to feed, the internal cells must get enough food and oxygen, and all the problems that attend increase in size must be met as they arise. For instance, a circulatory system is needed to pump oxygen and food to inner cells when the embryo thickens, and it is no wonder that in vertebrate embryos the heart is the first organ to start functioning. But I am not concerned here with differentiation, for that is a matter to be examined later in the discussion of complexity. Here the question is, how can large masses of soft cells, devoid of cell walls, produce a large animal?

The animal embryo develops through a combination of growth by the multiplication of cells and by the organized, directed movement of those cells that leads to the final form. In embryos the development must be adaptive, or at least not harmful, at all times. In some forms, the embryos are larvae, and it is easy to see how natural selection could affect the course of their construction steps; such is the case for sponges and cnidarians. In the embryos of more advanced animals it is the sequence of construction that becomes paramount. Even under the most protected circumstances the construction steps must be viable; if they are not, they will be selected out. So each step from gastrulation onwards has to be mechanically feasible. It has to be that all the cells have oxygen and food energy, and that they remain attached together by cell adhesion in an appropriate manner. As when building a house, the mechanics of the construction severely restrict what sequence of events is possible. There are different modes of gastrulation, which are possible because each one obviously works; but if a new one arose by mutation that was mechanically impossible for the subsequent developmental steps, it would instantly be eliminated. The selection pressure for sound rules of construction are strong and immediate; there may be many variations of any particular step in the building of an animal, but each of those variations must stand up to the scrutiny of mechanical feasibility. This is what L. L. Whyte (1965) has called "internal selection."

THE LEGACY OF PAST DEVELOPMENTS

This brings us to the final and exceedingly important consequence of the effect of size on development. If an animal or a plant is large, then, as I have just emphasized, it must involve many sequential steps in the construction process that we call development. Since those steps are complex and the success of each is dependent upon the ones that precede it, it is not surprising that during the course of evolution there should be a great conservatism in these steps. Once established, they have been, in their essence, left unchanged. As a result, if there is an ancestral plan of construction that works, it is, not surprisingly, found essentially intact in the descendants. Let me give some examples.

The best known is the vertebrate body plan. The laying down of the gut, the formation of the neural tube, the notochord, as well as the formation of the muscle somites are common features in all vertebrates. If one looks for finer detail, the transition from fins to limbs—and then ultimately the basic plan of four limbs each with five digits—emerges as an evolutionary sequence in which each step is built on the foundations laid down by previous structures. The same holds for the origin of the lungs from the air sac of fishes. Even when a structure is no longer needed, such as gills in terrestrial vertebrates, reptiles, birds, and mammals, it does not completely disappear but remains in the embryo in a nonfunctional state. The same is true for tails in birds and mammals. In both cases there is every reason to believe that the early stages have become part of the construction program needed for large animals, and there is no way of easily eliminating these now useless structures. There is certainly no selection pressure to remove them, and therefore it must be easy for them to be retained as harmless byproducts of a long and complicated construction process.

A similar story, but of a much simpler nature, is found in higher plants. The beginnings of the formation of meristems in the seedlings of all vascular plants, which include vast numbers of species, is basically the same. What differences exist are related to size. In the earlier stages of germination of seeds there will be intercalary meristems—that is, growth zones in the middle of a stem or root rather than at the apices—and these intercalary meristems are responsible for the elongation of the plant at the beginning stages. But as the size of the seedling increases and stiffer support is needed for the plant to rise into the air, the intercalary meristem disappears, and all the growth takes place at the apices of the shoot and root and all their branches. The only plants that retain their intercalary meristems are ones of negligible size, such as slender grasses. This is the reason we need to cut the lawn so often; the grass leaves grow at the base, not at the top. (It is interesting to note that this kind of leaf growth in grasses is undoubtedly a special adaptation to grazing by herbivores, but it is possible only by virtue of the small size of the plants.)

Even the construction of the meristem, which causes the thickening of the plant, is affected by size. In herbaceous plants the vascular tissue will be in discrete bundles separated by simple parenchyma cells, while in larger, woody plants the cambium becomes a continuous cylinder and can give rise to the thick wood (Fig. 41). Therefore, the largest trees go through a developmental sequence of three separate kinds of meristematic growth: (1) an intercalary meristem in the early seedling which soon vanishes; (2) apical meristems that persist indefinitely at each of the shoot and root tips; and (3) a persistent cylindrical cambium responsible for the increase in girth of the shoots and roots. The cambium first arises as separate bundles which ultimately fuse into a cylinder.

In this example there is no equivalent to gills or tails; no embryonic structures are leftovers from ancestors and no longer of any direct use to the organism. All the meristems are useful in both ancestor and descendant, and the only reminder of the evolutionary sequence from small to large is the sequence and mode of appearance of the three kinds of meristems. The construction plan is ancient in origin and has remained essentially unaltered since the beginnings of the evolution of vascular plants some 400 million years ago.

The very same arguments of conservatism of construction will be found in all groups of organisms that have achieved any size. For instance, the early steps of construction of large mushrooms are the same as those of small molds (Figs. 28, 29). If we turn to examples not previously mentioned, there are excellent ones in every major group of invertebrates. This is true of arthropods, not only in how they lay down their basic body plan or organs, but also in how they build and modify their exoskeleton. The largest, most elaborate arthropods build on the basic plan invented by the smallest and most primitive forms.

These very problems of construction have led to developmental con-

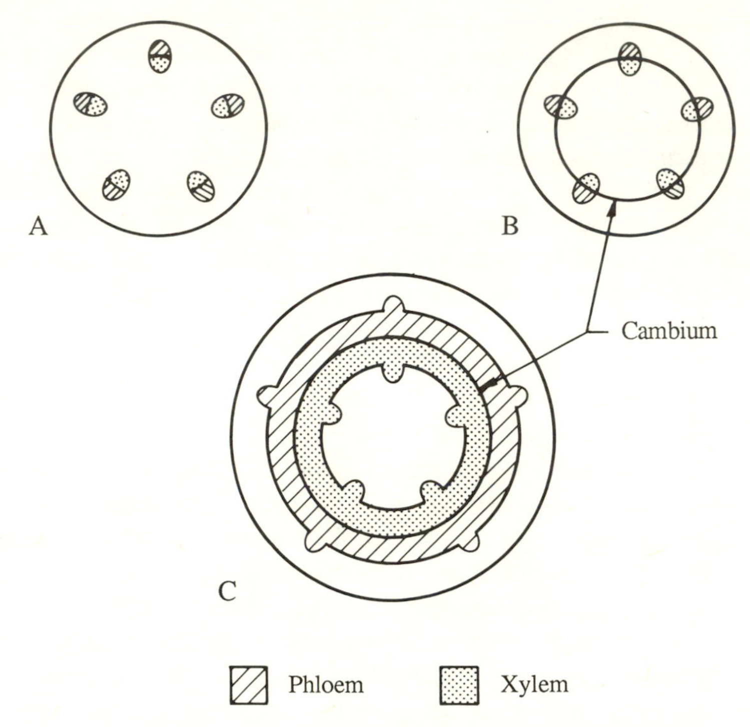

Fig. 41. Diagrammatic sections through the stem tip of a woody plant showing that near the apex (*A*) there are separate vascular bundles as in herbaceous plants, but farther down the stem the lateral growth zone, or cambium, becomes a continuous ring or cylinder (*B*) and is responsible for the thickening of the plant (*C*). (After Raven et al., *Biology of Plants*, 1986.)

servatism and have provided the basis for all the discussion and speculation on the subject of recapitulation in both its palatable and unpalatable forms. Strictly speaking, recapitulation, in the sense that the ontogeny of an individual recapitulates its evolutionary or phylogenetic history, means that during development one sees the sequence of that individual's adult ancestors. As S. J. Gould (1977) indicates, this simplified form of the relation of ontogeny to phylogeny has long been considered totally inadequate, beginning with K. E. von Baer who, in the early part of the nineteenth century, was the first to make some important and lasting points about the phenomenon. He made clear in his famous laws of development that the embryos of vertebrates more closely resemble one another than their respective adults. The adults diverged and were not, as adults, recapitulated in the embryonic stages of descendants; rather, the embryos from different groups of animals were repeated. Note that this is precisely what one might expect from the argument given in this chapter—that in any one group of organisms, as evolution proceeded from small to large, the early construction steps of the embryo would be retained in the descendants. So the explanation of von Baer's law that embryos of different species within a group of organisms tend to resemble one another has its explanation in the difficulty of altering the sequence of construction steps in large developing organisms.

Further References

I have previously discussed various aspects of the origin of multicellularity and the period of size increase in development (Bonner 1952, 1958, 1965, 1974). For a critical discussion of the elastic similarity model, see K. Schmidt-Nielsen (1983) and W. A Calder (1984). The matter of growing trees and the work of H. S. Horn and his associates are briefly reviewed in T. A. McMahon and J. T. Bonner (1983: 144) and discussed in greater detail in a field biology course manual prepared by H. S. Horn. Our understanding of inversion in *Volvox* comes largely through the work of J. L. Kelland (1977) and G. Viamontes et al. (1979).

Chapter 5

Relation between the Complexity of Communities and the Size, Diversity, and Abundance of the Organisms within Them

Complexity ▪ Diversity and size ▪ Diversity, abundance, and size ▪ Diversity and habitat ▪ Evolution of diversity

Complexity

The key message of this book concerns the relation between size and complexity. Before going into that message it may be helpful to say what I mean by "complexity." We have been spoiled by the earlier discussion on size, which is so easy to define; complexity presents a more difficult problem.

H. A. Simon (1962) considered complexity, or complex systems, to be made up of a large number of parts that interact, or, as the dictionary says, that are connected together. This immediately makes the point that there are many parts, but I would like to go further than merely to say the parts affect one another. It seems to me the most interesting thing is that parts are often not only numerous, but frequently they are different in their structure and function. As we shall see, usually this aspect of complexity gives a division of labor; in fact, this kind of interaction of parts is most characteristic of complex biological systems.

It is also standard to think of complexity as being hierarchical, and Simon makes the important point that such hierarchical structure leads to efficiency of construction. Each part has subparts, and each subpart may be further divided. An obvious example is the mammal, which is made up of organs, which are in turn made up of cells, then molecules, atoms, and finally elementary particles. On the other end of the scale are populations and societies.

Here I want to emphasize that these hierarchical levels are related to size. Populations involve many individual organisms, and all the internal divisions of the body are obviously smaller than the body itself. Therefore the notion that hierarchical compexity is size-related is self-evident perhaps so much so that it is rarely emphasized. Here the particular interest is in size, and this makes its relation to complexity especially important.

While I have given a general definition of complexity, it is necessary to be more explicit on how the word will be used here. In the first place, I will be examining only two levels in the hierarchical chain, that of populations and that of the parts within an individual organism; the complexity to be examined here will be purely biological.

If one looks at the complexity within organisms, then one can equate the subdivision of the innards of the body with a division of labor. There are cells of tissues or organs that specialize in specific functions such as gas exchange, or waste removal, or energy intake, or locomotion, or coordination, and these can be identified by their particular morphologies, which seem to be admirably suited to perform their share of the labor. In vertebrates the cells of the lung or the gill are specialists in getting atmospheric oxygen to the blood; the tubules of the kidney selectively eliminate unwanted chemicals; food, or energy, is assimilated through a gut; movement is achieved by muscular contraction, and those contractions are coordinated in the whole body by the nervous system. We could go on in great detail, but what we need now is simply to understand the concept. I also want to make the point that probably the easiest way to consider and measure all this division of labor is to count the number of different cell types involved for any particular animal or plant and see how this number correlates with size. If one attempts to include other levels, such as organ systems, organs, and tissues, it becomes difficult to compare different kinds of animals and plants; cells are the easiest common denominator. The only qualification that should be made to the above is for those cells that appear to be identical in structure, yet serve different functions. The example of such a phenomenon that will be encountered later is the nervous system. In that case vast numbers of neurons that seem to be similar morphologically can function in a variety of different ways, thereby greatly increasing the organism's complexity in the form of its behavior.

Let us now consider complexity at the population level. Here one is concerned with communities of organisms in different environments, aquatic or terrestrial. Size in this case, as was pointed out previously, is reflected by the total number of organisms in that community, and within that total number there are many organisms of different size classes. If we now consider any community in terms of complexity, we must again seek some form of division of labor. Ecologists often think of a community as a collection of niches, and these niches reflect different ways animals or plants can maintain a place in the community. This is hardly identical to the kind of division of labor within the body in the form of cells and tissues, but it is a close parallel. If one thinks of niches as slots where organisms can lead a different kind of existence, functioning and behaving in specialized ways, then, if it is not a strict division of labor, it is a division of ways of living. Furthermore, it is often true that the way one type of organism lives is a necessary requirement for the existence of another type, a kind of interdependence clearly analogous to that of the parts within the body of an individual organism. This not only applies in an obvious way to all parasites and obligate symbionts, but in larger ways: for example, all the animals within a community are ultimately dependent on the photosynthetic plants, as the carnivores are dependent on the existence of other animals, and by the same token an epiphyte, such as an orchid, depends on a large tree to hold it up into the air, and so forth.

If one looks for differences in closely related species in a community, one seeks differences in their patterns of living that are sufficient to isolate them from one another so that they do not normally cross-breed. Isolation at this most fine-grained level may be due to feeding preferences, that is, to a subdivision of food resources that allows closely related species to specialize in different kinds of food. For instance, as discussed earlier, in birds some finches eat relatively large seeds, and a closely related species specializes in smaller seeds. The specialization that separates species may not necessarily involve food, but it could involve the occupation of different zones in the habitat which might differ in their diurnal characteristic temperatures, or humidities, or in other physical characteristics. New species seem to form in such ways that they lead either to the subdivision of a former niche, as in the case of seed-eating birds, or they invade a new niche that had not previously been occupied at all. We discussed an example of this earlier, where an increase in the size of very large organisms can result in the avoidance of competition.

These ecological comments are meant to make the point that the complexity of a community can be measured in terms of the number of species that exist in it. For a whole community, this is the simplest and most realistic equivalent to the use of cell types as a measure of complexity for an individual organism. The number of species in any one place is called species diversity, or simply diversity. Therefore, in the world within an organism, complexity will be measured by the number of cell types, and in the world that surrounds an organism, that is, the world within which it lies, complexity will be measured by the species diversity. For the discussion here, this will be our straightforward, operational definition of complexity. The greater the number of cell types, or the number of species, the greater the complexity.

These two levels of complexity are totally biological, but one might well ask whether there is any intrinsic difference between them and the hierarchies of the physical world. How do they differ from the complexities of the heavens, or those of molecules and atoms? There is an important difference: stars and molecules are solely governed by physical laws, while populations and individual organisms are governed by natural selection, which steers a genetic system, albeit within the constraints of physical possibility. Living organisms possess genes which are directly inherited by their offspring, and the success of the individual in its ability to reproduce determines which genes prevail in subsequent generations. The changes in genes are by chance, such as mutation, and their preservation is either again by chance (as in genetic drift) or more often by the culling of natural selection. Since one cannot have selection without reproduction in successive generations and a system of preserving the changes by inheritance, natural selection is a process that can be managed only by living organisms. It does not defy any laws of physics; it is a purely biological phenomenon.

Diversity and Size

In Chapter 2 it was emphasized that during the course of evolution the size of organisms increased; there was a progression of upper size limits with time. Implicit in that trend is a concomitant increase in the complexity within the individual organisms of increasing size, a subject to be discussed in the next chapter. Here we will examine the changes in whole communities over evolutionary time, and again we find that their complexity increases.

If one takes some simple examples the trend is easy to see. Presumably at some very early stage in earth history the only existing organisms were bacteria. At a later stage there must have been a period in the Precambrian (that is, some time earlier than about 700 million years ago; Fig. 22) when a few small invertebrates and some algae coexisted in a community along with protozoa and bacteria. Finally, consider a community today, either on land or in the ocean, where there are organisms ranging in size from bacteria to large trees and mammals, or to huge kelp, large fish, and whales. If one measures complexity in terms of the number of species, clearly these three periods in earth history show a substantial increase in the complexity of biological communities.

We can understand this evolutionary increase in complexity more readily if we look at the relation between the size of a group of animals or plants and their diversity. This is a subject that was first examined quantitatively by G. E. Hutchinson and R. H. MacArthur (1959) and later by others, including R. M. May (1978), who presents a clear review of the earlier work. The important point is summarized in May's graph which shows how the number of species of terrestrial animals varies with their size class (Fig. 42). The data for this curve come from a large number of sources and span a range from small insects to large mammals. Note that going from big animals down to ones that are a few millimeters long, the curve rises: relatively small animals are characterized by more species than large ones for most of the size ranges. The dotted line has a slope of -2, but May cautions against attaching any great significance to this approximate figure; it should be thought of only as a rough guide.

Before going further, we must ask why the curve peaks and falls off for very small species in Figure 42. How can we explain this change? As is so often the case, there is probably more than one explanation. Let me give some examples from different groups of organisms. For instance, this is true for birds (Fig. 43), and S. M. Stanley (1973) has argued that there are mechanical limitations at the lower end of the size scale for animals with a bird body plan, and that these limitations have been overcome by relatively few species. In the case of small birds, the metabolic cost becomes very great at sizes below 4 to 5 grams. Even the heart has difficulties, for if it were scaled normally to body size it would beat at an impossibly rapid rate, and for this reason hummingbird hearts are relatively large. In many of these instances, one might postulate that the peak in the size-diversity curve is the optimum size for a particular level of complexity.

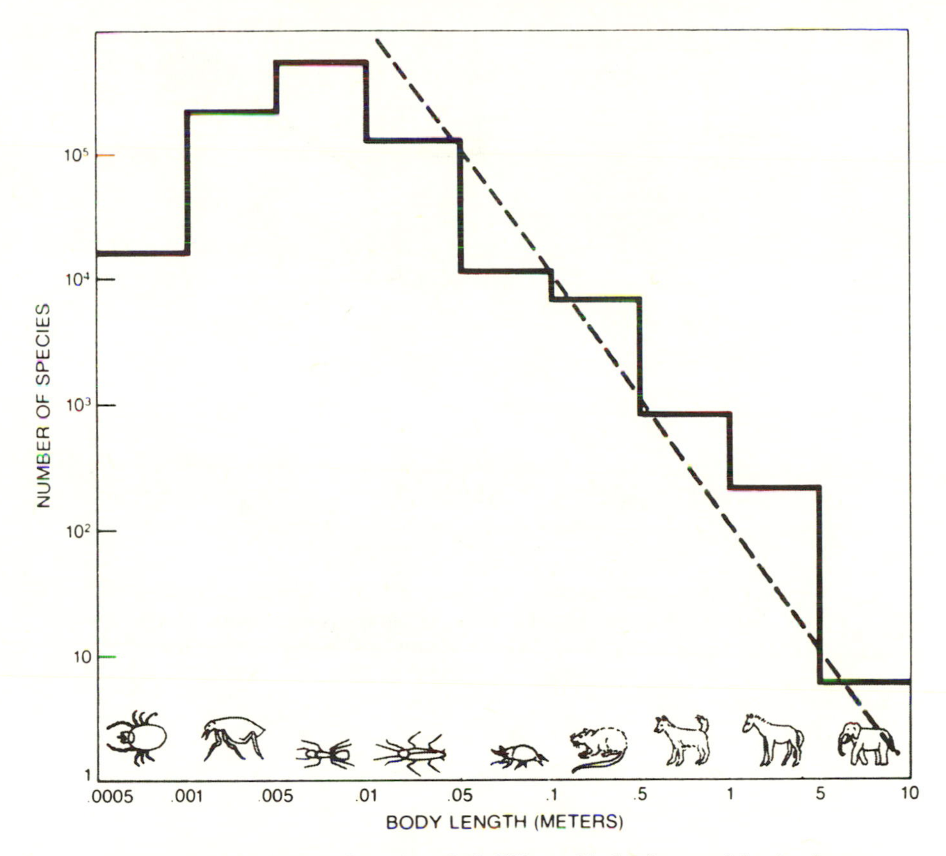

Fig. 42. The number of species of all different kinds of terrestrial animals, grouped according to length on a logarithmic scale. (From May, in *Diversity of Insect Faunas*, Mound and Waloff, eds., 1984.)

The same drop on the left-hand, "small" end of the curve may also be found in insects (Fig. 44). Here too the minute forms could be reacting to some sort of physiological limit that is difficult to attain. But there are other possibilities: for instance, finding them and classifying them is a more difficult task than doing the same for larger ones. This would mean that a possible reason for some of the deficiency of small species of insects is a reflection of the problems of practicing taxonomists, although I suspect they would contest this point.

None of the arguments apply in the case of trees that show the same size-diversity relations (Fig. 45). Here it is unlikely that small trees have any kind of physiological limit, and certainly they are not difficult for the plant taxonomist to find. In this instance I wonder if the reasons

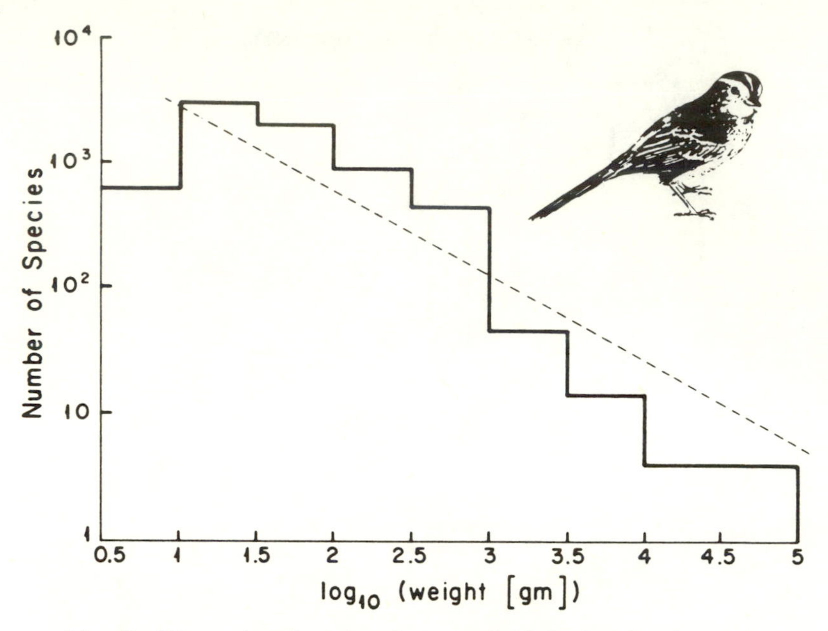

Fig. 43. The number of species of nonaquatic birds plotted against their size, based on different weight classes. The graph is logarithmic. (From May, in *Diversity of Insect Faunas*, Mound and Waloff, eds., 1984, based on data of Van Valen 1973.)

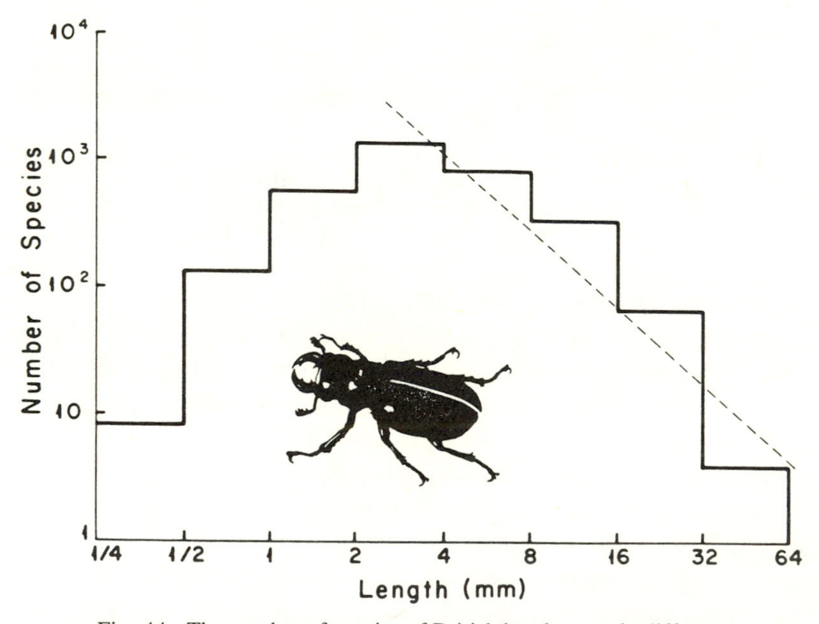

Fig. 44. The number of species of British beetles, put in different weight classes. (From May, in *Diversity of Insect Faunas*, Mound and Waloff, eds., 1984, largely based on data of Fowler 1887.)

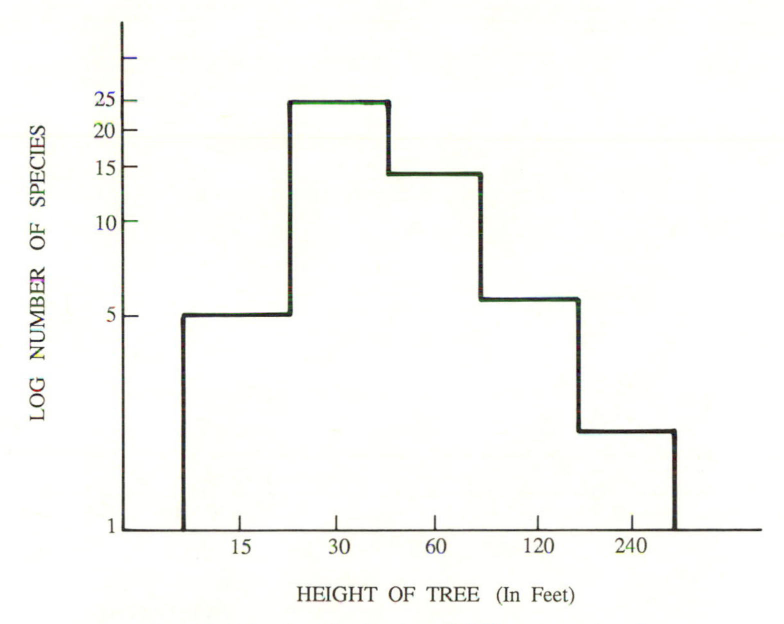

Fig. 45. The size distribution of different species of mature North American trees. (Data from Brockman, *Trees of North America*, 1968.)

for the paucity of small tree species might be important and deserve special attention.

It is particularly interesting that bacteria and other prokaryotes exhibit the same kind of size-diversity relationship, as has been shown by H. J. Hofmann and J. W. Schopf (1983) (Fig. 46). Even though the total number of species is small, the curves show a peak of diversity in an intermediate size range for both bacteria and cyanobacteria. Therefore this would appear to be a general rule for all taxonomic groups regardless of their size.

If all organisms are included—that is, the entire size span from bacteria to the largest animals and plants—there is the same kind of curve, with reduced number of species at both the large and small ends of the scale (Fig. 47). This implies that there are fewer species of prokaryotes than many of the major groups of eukaryotes, and this indeed seems to be the case. However, there is a special problem of identifying species among microbes. Certainly the criteria for what constitutes a species in bacteria and in higher animals and plants are very different and hardly comparable. Even among different groups of large organisms, the cri-

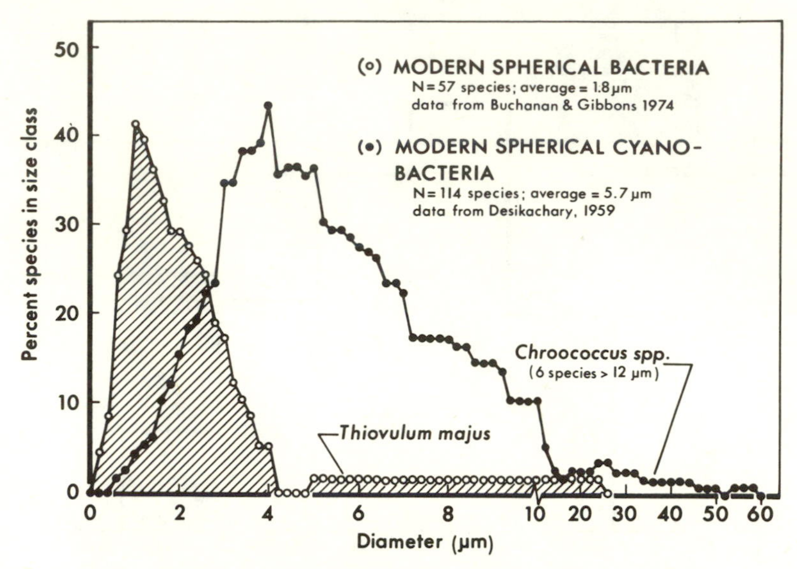

Fig. 46. The size distribution of modern spherical bacteria and cyanobacteria. (Note that this is a linear plot of percent species in each size class, plotted against the diameter of the organisms.) (From Hofmann and Schopf 1983, Copyright © by Princeton University Press.)

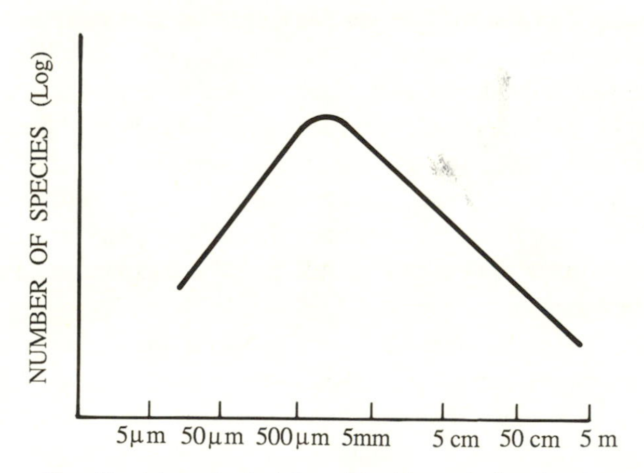

Fig. 47. A hypothetical (logarithmic) graph showing the relation between the size of all organisms and the number of species.

teria are sometimes hard to compare, and the difficulty becomes enormous when one descends into the microworld. It seems to me there are various possibilities, some of which have an air of fantasy to them. One is that there is no equivalent to the species of larger animals and plants among microbes. Another is that we simply do not understand the richness of the microbial world and how microbes subdivide their environment into niches. Perhaps a more sensible expanation is that their world has very different physical properties from those of large forms, and this might have an influence. The forces of diffusion, cohesion, surface tension and viscosity are all exceedingly high in the microenvironment of bacteria, while gravity and inertia are dominant for large animals and plants. But this surely cannot be the whole reason for the small number of microbial species; more likely it is a combination of some of the factors we have mentioned, and no doubt of others that have not yet been imagined. And there is always the possibility that in many instances the peak of the size-diversity curve is an optimum size for a given group of organisms.

It should be pointed out, before concluding this discussion, that these diversity-size curves are based on all the species known usually in a major geographic area, and not on the species found in a single ecological community. The assumption is that the curves would apply equally well to any one community, and it is a reasonable assumption. What is needed now is a complete analysis of different kinds of communities in which all the animals and plants are measured, both for their size and for their abundance, and classified by species. However, a more difficult and laborious plan could scarcely be conceived, and it may be a long time before we have such facts and figures.

Diversity, abundance, and size

We can also link diversity with abundance because they are both related to size. In Chapter 3 we saw that the abundance or density of an organism is roughly inversely proportional to its size (Fig. 16), the very same rough relation that is true for diversity and size (Figs. 42 to 47). This means that more abundant classes of animals and plants, that is, those that exist in the highest density, will also have the highest diversity. This relationship between abundance and diversity can be seen diagrammatically in Fig. 48.

There is, therefore, a very general correlation between size, abun-

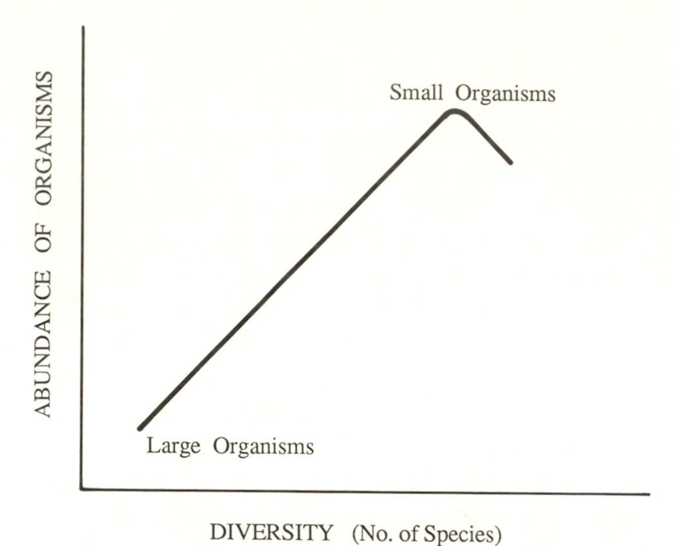

Fig. 48. A hypothetical graph showing that in any one habitat, when there are more species, those species are represented by more, but smaller, individuals.

dance of individuals, and the number of species. As with all correlations, one immediately wants to understand their basis: which of these elements is the driving force, or more specifically, which of them can be affected by natural selection.

Abundance is the least likely candidate, for it would appear to be determined solely by the energy available for a particular size class. Or put another way, if the energy available for each size class is the same, then so will the biomass. Density is simply determined by dividing the total mass of a size group by the average size of an individual in that particular size class, and the smaller the individual's size the more numerous it will be in a given community.

We have already seen that size, by itself, may be influenced by selection; now let me ask if this is also true for diversity. The best way to approach the question is to think of diversity in terms of morphological distinctions between species, and therefore the question can be rephrased: Is there a selection for shape? The answer is obviously yes, but the more interesting problem is not whether such selection exists, but what is the relation, during the course of evolution, of selection for size and selection for shape. Since changes in shape involve development, this is a subject that will be carefully examined in the next chapter. Here let me say that, in general, radical shape changes occur most effectively in the smaller species of a given group of animals and plants, and that size changes usually follow the shape changes. This is the case, for instance, in the origin of many major groups of animals such as mammals, or plants such as vascular trees, as was discussed in Chapter 2. The innovations that led to a major group of structurally different organisms (vascular plants or vertebrate animals) all occurred in small, ancestral forms, and the changes were followed by a great increase in size as a result of selection. The size increase was accompanied by only relatively minor selection for changes in shape. So selection for shape occurs all the time, but the most fundamental changes tend to occur in small members of a taxon.

This does not mean that major morphological changes always precede changes in size increase; let me give a clear example of the reverse. The cellular slime molds are characterized by the chemotactic aggregation of single amoebae to central collection points, to form a multicellular organism. The collected cell mass differentiates into spore and stalk cells; this is the division of its internal labor (Fig. 30). As discussed in the previous chapter, in one species there is no such division of labor and the cellulose stalk is acellular. All the cells secrete the stalk, after which all the cells turn into spores. We have no way of knowing if *Acytostelium*, as it is appropriately called, is ancestral, but it shows that slime molds are possible with and without a division of labor. In Chapter 4 I argued that there is evidence for a selection pressure, presumably to enhance spore dispersal, for fruiting bodies that are about a millimeter high. If that is so, then the first step in the evolution of these social amoebae must have been aggregation; this is the way they increased their size. Perhaps long after this first evolutionary step took place, the division of labor in the form of stalk cells and spores occurred. It may well be that the ancestral aggregate in no way resembled *Acytostelium*, but then again it may have had a similar construction. In fact, one finds that the acrasids, which are a group of soil amoebae distinct from these dictyostelids and undoubtedly had a separate origin because their cell structure is very different, also produce small fruiting bodies with and without a division of stalk and spore cells. We could conclude that these early attempts at multicellularity arose first in response to a selection pressure for increase in size, and this was followed by selection for a change in shape accompanying a division of labor.

If we turn to all those multicellular animals and plants that come from a single-cell fertilized egg or a zygote that enlarges by successive cell divisions, we can ask again which came first: size increase or change in shape of the constituent cells. The best way to look at this problem is to consider some of the simple colonial organisms, such as the *Volvox* line of green algae. Here in the smaller forms the cells are identical, and it is only in large colonies, the size of *Volvox* itself, that there is a separation of vegetative cells and reproductive cells (Fig. 25). One could find many more examples, but the point is similar to that made for slime molds. The first steps towards multicellularity, whether by cell aggregation or by growth of a single cell into numerous, attached daughter cells, all involved a size increase. Any division of labor or major change in shape is a secondary refinement. So we see that important changes can begin either as size changes or shape changes, and less fundamental changes in shape and size are constantly being produced separately during evolution.

I must remind the reader that shape is not quite the same thing as diversity, although it has been useful to equate the two for this discussion. It is also possible to make two species out of one by altering only the size of the individuals in the two populations so that they ultimately differ significantly. But it is true that species diversity is more often reflected in differences in shape and in construction, and that is the justification for considering them together here.

DIVERSITY AND HABITAT

One of the important themes in modern ecology is the relation of species diversity to the habitat, which includes the physical environment and the richness of energy in the form of available food resources. The first big stimulus came from G. E. Hutchinson's famous essay "Homage to Santa Rosalia, or why are there so many kinds of animals?" (1959), and since then a large literature has emerged. This is not the place to write a review of this subject, nor even a textbook survey, but rather I will glean a few points from the extensive discussion that seem especially pertinent to the argument here. We are concerned with complexity at the ecological level, and clearly whatever factors affect the number of species in any one community are relevant, especially since we have equated complexity with species diversity.

The most obvious point is that if we compare a portion of a tropical rain forest with that of a temperate forest of approximately the same amount of total energy production (or biomass), we will see many more species in the tropical habitat than in the temperate one (and an arctic habitat will have even fewer species). If the productivity is roughly the same, then, since there are so many more species in the tropics, on the average there will be fewer individuals of each species compared to the species of the temperate forest. Indeed, for many groups of organisms which have been measured—birds, mammals, marine molluscs, insects, plants (although there is some controversy here), and even slime molds—there is a latitudinal gradient which shows that as one moves towards the equator, the number of species for any particular group increases.

The principal reason for this relation appears to be the stability or the predictability of the warmer environment. Closer to the poles, the seasons become more severe and more marked, and the suddenness with which a great freeze or a thaw can hit provides an environment that is one of continuous hazard. The same phenomenon is seen on a small scale on mountains, where with increasing altitude the environment becomes harsher and less predictable.

In general, it is clear that changeable environments favor those organisms which produce many offspring and have short generations (that is, are *r*-selected), while steady environments favor organisms with few offspring that have a good chance of survival (namely, are *K*-selected). *r* selection is ideally suited to overcome sudden environmental catastrophies, because the few young, or seeds, or dormant eggs that survive can repopulate the community rapidly and take advantage of abundant food resources. On the other hand, *K* selection is favored in circumstances where there is intense competition among species, and those that manage to place their offspring in the few slots that are available in the tough world they enter will survive. The relation of *r* and *K* selection to size has already been discussed in Chapter 3, and I mentioned that to some degree the concept was an oversimplification, but one of such usefulness that its advantages far outweighed its shortcomings. Let us now examine the concept with relation to habitat; as will be evident, it is also closely correlated with complexity.

Tropical forests have a large number of species of animals and plants, and within any one group of organisms there is a predominance of relatively *K*-selected species. This means that the competition is intense between the species to place offspring successfully so that they can reproduce. Furthermore, because there are so many species, albeit each in small numbers, there will be a very large number of interactions among those species. The niches are so similar and overlap to such a degree that the most important kind of interaction will be between the similar, competing species themselves. But more species also mean more predators and therefore a large increase in the interaction between all the different kinds of prey and the predator, as well as many patterns of mutualism and parasitism. Another way the complexity is increased in a tropical forest is by the very tall trees that permit a layering of subcommunities from the top of the canopy to below the surface of the soil. Each level will have its separate structure, its separate space, and often its separate animals, microbes, and epiphytic plants such as orchids and bromeliads. Again this structure in the community is largely the result of the predictability or stability of the climate. The result is an extraordinary amount of complexity in the spacing of the species and the interactions between them.

In the temperate forest there is no difference in principle of how the community is put together; the difference is purely a matter of degree. There are fewer species, less structure, and therefore fewer interactions involving competition, predation, mutualism, and so forth, and as a result there is less complexity.

In this brief and oversimplified statement of the case for habitat complexity, there are too many things left unsaid, so let me add a few points. One is that the initial work on diversity in different habitats was done by zoologists, and the case for plants is undoubtedly different. There are indeed more species of trees in the tropics than in a temperate forest, but it is difficult to use the same arguments we used above for animals to explain this difference. Recently D. Tilman (1982) has shown that although plants all seem to need pretty much the same resources—sunlight, carbon dioxide, and various nutrients in the soil such as nitrates, phosphates, and other inorganic compounds—they often exhibit great diversity in small plots of land. He has discovered that these soil nutrients are not evenly distributed but vary widely over short distances, and the plants have strong preferences for certain optimum concentrations of particular nutrients and are especially sensitive to the ratio between nutrients in the soil. This means that as far as a plant is concerned, the soil itself is a complex pattern of different combinations of nutrients, and each species of plant has concentration and ratio preferences. So plants compete for nutrients, but the question of who wins or loses is predetermined in the soil. In this way it is possible for different species of plants to live in close proximity and therefore show high diversity in a fashion very similar to what one finds for animal diversity. The key to success is not in how much nutrient there is, since the optimum for the plants is usually relatively low. In fact, Tilman points out that by fertilizing the soil one can decrease the number of species that it will support. He uses this very argument to explain the great plant diversity in the tropics; soils in the tropics are less rich than temperate soils, and this favors diversity. Tilman makes a very strong case and, although it may not be the final word on plant diversity, I use it here as an example of the kind of detail one wishes to have on the causes of the great diversity of communities such as a tropical forest.

There is also one puzzle that especially interests me. Why should soil amoebae show a latitudinal trend, so that if one travels from Alaska to the tropics the number of species increases (Fig. 49)? Perhaps it is that the food organisms are more diverse in the tropics than farther north. E. G. Horn (1971) showed that the coexistence of different species of soil amoebae in a small plot of soil could best be explained by their preference for eating specific strains of bacteria that coexist in the soil. But this would mean that there are more "species" of bacteria in tropical soils than soils farther north, something that has not been estab-

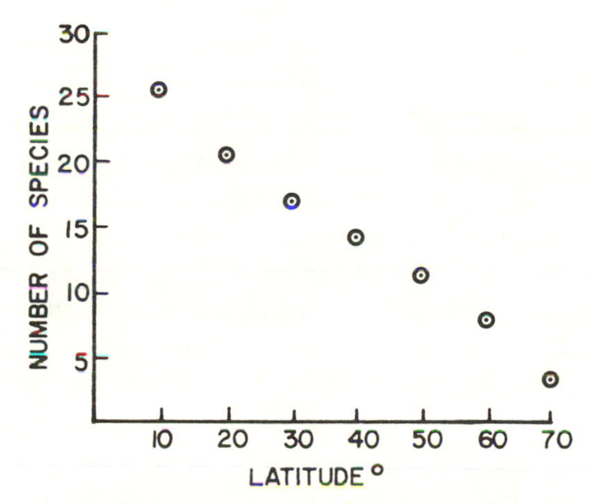

Fig. 49. The number of cellular slime mold species at different latitudes. There is more species diversity at the equator than in the arctic. (From Cavender, *Can. J. Bot.* 56:1326–1332 1978.)

lished, although, as my colleague Henry Horn has pointed out to me, it may be that the bacterial food in tropical soil is less abundant and distributed in patches, which would lead to diversity. There are many significant and intriguing problems yet to be solved in the ecology of diversity.

Evolution of diversity

It is now timely to return to a theme I discussed briefly at the beginning of this chapter, namely the increase of diversity during the course of evolution. There have been innumerable discussions in the literature of how one species might evolve into two by means of various isolating mechanisms, but here I am asking a bigger question: How is it that the total number of species increased over geological time, and how did the new niches arise?

The principle is straightforward: it is by avoiding or succeeding in competition or predation that newly formed species will find new niches. In Chapter 3 this was called "pioneering," and I pointed out that one important and continuous bit of pioneering has been the evolution of larger animals and plants that ultimately avoid or succeed in competition and predation by being big. Let us now examine first the simplest kind of community and see how pioneering could lead to increased diversity.

In primeval mud, which contained no organisms other than bacteria, the first limiting factor was bound to be the source of energy. Today we know that some prokaryotes use the sun's energy and are photosynthetic, while others get their energy by encouraging a simple chemical reaction such as the conversion of nitrates to nitrites, or sulphur to sulphur dioxide, or iron to ferrous oxide, or even hydrogen into water and other similar chemical processes. Each of these reactions gives off energy, and these so-called chemosynthetic bacteria grab that energy and use it to build their own proteins, carbohydrates, and other organic cell constituents. Finally, many bacteria can directly use organic substances, such as sugars in the surroundings, but this must have been a relatively late evolutionary development.

Each of these profoundly different ways of obtaining energy is an innovation, a bit of pioneering. By inventing any one of these methods of capturing energy, the bacteria can avoid competition; they make a new world of their own. What we do not know is the historical relationships between the chemosynthetic and the photosynthetic prokaryotes. For our purposes it is only important to think of them as a proliferation of energy-catching techniques. We can assume that initially all prokaryotes used the same method, whatever that ancestral method might have been, and that as competition for the first substrate increased, new energy-trapping inventions were devised. It is more difficult to see how photosynthesis could compete with chemosynthesis, except that again there are two worlds: places where the sun reaches the surface, and places deep in the mud where it cannot.

This is not the place to discuss how the eukaryotic cell arose from the earlier prokaryotes, an important subject which has stimulated much interesting speculation. Whatever the steps, and whatever the forces of selection, it produced diversity in the form of a new type of organism with a radically different kind of cell plan, one that has been as extraordinarily successful and stable as prokaryote cell construction. One of the important new things about eukaryotic cells is that, besides being able to maintain the ability to perform photosynthesis and take in dissolved foods such as sugars directly, they also invented a new method of capturing energy. It is the engulfing of particulate food; early eukaryotic cells could eat and digest bacteria, thereby becoming predators. The appearance of eukaryotes provided new opportunities for prokaryotes too. Some species of bacteria learned to grow inside eukaryotes and to become symbionts with mutual benefits; in other instances the association is one-sided, that is they became parasites. Each of these steps is pioneering; each means an increase in diversity.

Once predation was established, there must have been a whole new set of selection pressures on the bacteria, for there would be strong selection for predator avoidance on the part of the bacteria. This could well have led to still more diversity.

The next major step in evolution might have been the invention of multicellularity. In addition to the diverse unicellular bacteria, there evolved the larger prokaryotes, the actinomycetes, the blue-green algae, and many different colonial eukaryotic animals and plants. This innovation of size increase generated all sorts of new possible interactions, involving both competition and predation, and a new element appeared: dispersal became an effective means of persisting and competing. In the case of many mobile forms, larger colonies can move faster than smaller ones and thereby find new favorable patches for further propagation. In other primitive multicellular forms dispersal is apparently helped by masses of spores being bunched together, a matter discussed in Chapter 4.

The advantages of multicellularity in predation are especially interesting. Various authors have suggested that the multicellular swarms of bacterial cells in the myxobacteria, and the multinucleate feeding plasmodia of myxomycetes, or true slime molds, are so successful because both secrete massive amounts of extracellular enzymes that dissolve or break down large food particles, making available the simple sugars and amino acids. They are able to do this because they are large and therefore produce high concentrations of enzymes that can break down organic matter which would be unavailable to small, single, feeding cells. Another solution to predation is found in sponges that feed on small particulate food; by creating strong currents they bring in fresh water laden with food and oxygen. Finally, in animals there is the important invention of the gut—first with one opening, as in cnidaria and flatworms, and then with an entrance and exit, as in higher animals. I need not emphasize here all the mechanical advantages and design features of these remarkable innovations, but instead I will stress how each leads to the formation of new species.

A final point about pioneering concerns an organisms's entrance into an entirely new kind of habitat, the most conspicuous of which is the conquest of land or of the air. In each case the removal to a new environment was again designed to avoid competition and predation; it was an escape. It has happened many times during the course of evolution. For instance, there are terrestrial bacteria such as the myxobacteria, which make small fruiting bodies that rise up into the air; the same is true of amoebae in the form of slime molds. Terrestrial molds form a very large group, and there are even a number of terrestrial algae. In all of these examples the organism cannot be too far removed from moisture; but in the more complex mosses, ferns, and higher plants, special mechanisms to avoid desiccation have been invented, as well as ways to avoid the need of water for fertilization and sexual reproduction. The same trends are found in animals: many invertebrate groups—such as worms, molluscs, and, most conspicuous of all, insects and other anthropods—have devised wonderful tricks to manage effectively on land. Finally, the evolution of land-dwelling vertebrates provides an ideal illustration of the point, for again we see ways of preventing water loss and ways of managing fertilization without a water environment to carry the sperm to the egg.

In the conquest of air, insects, birds, and bats are the star performers, although one should not forget the flying reptiles of prehistoric times, or gliding spiders and squirrels. In each instance these separate conquests consisted of a dramatic escape from life on the earth; it permitted greater facility to escape predators, better opportunities to find food, and the opening of a new world.

However, we must remember that we are not concerned here with the extraordinary cleverness of the inventions, but rather that they led to the formation of many more new species. Each change in speciation led to further changes, so that diversity bred even greater diversity. We assume that such proliferation of species, as a result of these innovations, must reach some sort of saturation, but who knows what that limit in complexity might be? When all organisms were aquatic, it could have been assumed that saturation existed, but with the opening of land came a whole new wave, and the same is true for the conquest of air. The inventiveness of nature and the diversity that goes with it have been endlessly surprising in the past, and no doubt that tells us something about the future.

Further References

The question of mechanical limits of small size is discussed in K. Schmidt-Nielsen (1984) and in W. A. Calder (1984). There has been an extensive discussion in recent years on the origin of the eukaryotic cell. For reviews, see M. J. Carlile (1980) and L. Margulis (1981).

Chapter 6

How Size Affects the Internal Complexities of Organisms in Their Evolution and in Their Development

Clonal vs. aclonal organisms ▪ Size and internal complexity ▪ Cell size and internal complexity ▪ Genome size and internal complexity ▪ How is a large, complex organism built? ▪ Plasticity in development ▪ The evolution of internal complexity ▪ Modifiability and complexity ▪ Natural selection and development

IN THE last chapter the complexity of communities was examined; here I examine the complexity within individual organisms. The discussion is therefore shifting from one hierarchical level to another, from the problems that exist between organisms to those that exist within organisms.

CLONAL VS. ACLONAL ORGANISMS

It is important to stress that when I talk about how increases in size and complexity go hand in hand, I am referring exclusively to aclonal organisms—those animals and plants that reproduce each life cycle sexually. In clonal organisms, there is a great asexual proliferation of buds, and if the products of these buds all remain attached to the parent, there will be one huge colony of genetically identical individuals. In animals, connected clones are found among cnidarians such as corals and colonial hydroids, and they are common among bryozoans, an abundant, primitive group of encrusting marine invertebrates that is particularly successful in spreading over surfaces on the ocean floor. Clonal proliferation is also common among plants, and good examples may be seen among club mosses and many species of angiosperms, from small herbaceous plants to large trees such as the aspen. In these instances the entire colony may reach enormous size, although the many identical individuals that make it up remain comparatively small, and their complexity is commensurate with the size of the small individuals and not with that of the entire colony. Producing colonies of clones is an inexpensive way of becoming large; it is a dodge whereby the difficult problem of becoming complex during the course of evolution, which involves many genetic steps over great periods of time, can be sidestepped or avoided by the formation of clones.

A particularly revealing aspect of this problem has been a discovery by B. H. Tiffney and K. J. Niklas (1985). They examined the fossil record for plants and showed that from the Silurian (about 400 million years ago) to relatively recent times there has been a trend towards an increasing percentage of aclonal angiosperms. The first flowering plants were exclusively clonal in their mode of size increase, which did not involve any increase in complexity. Over many millions of years of evolution, clonal plants became a minority compared to the relatively more numerous, complex, and large aclonal ones.

The same is true for animals, and A. G. Coates and J.B.C. Jackson (1985) have found the same trend in the fossil record for certain corals. Furthermore, they show an additional, but related, important trend. With geological time the clones of corals have become more integrated. By this they mean that not only are the buds connected, but one also begins to see a division of labor, where some polyps do all the feeding while others specialize in protection or in reproduction. Moreover, these polyps may be in close communication with one another. The pinnacle of this kind of integration may be seen in the siphonophores, which are related to the colonial hydroids (Fig. 50). Siphonophores such as *Physalia* or the Portuguese man-o'-war and many smaller forms have evolved from a simple, clonal colony to a new super ''aclonal'' organism, in which the individuals, by their extreme division of labor, become subjugated into a new whole. By a trend in integration the colony of attached individuals has become a unified organism at a higher level.

One final word about clonal organisms. They not only achieve great size without a concomitant increase in complexity, but they also escape other restraints that seem to be connected with complexity. For instance, as J. A. Silander, Jr. (1985), has made clear, there is no neat correlation between size and life span in clonal organisms. In aclonal forms, in general, the larger the animal or plant, the greater its longev-

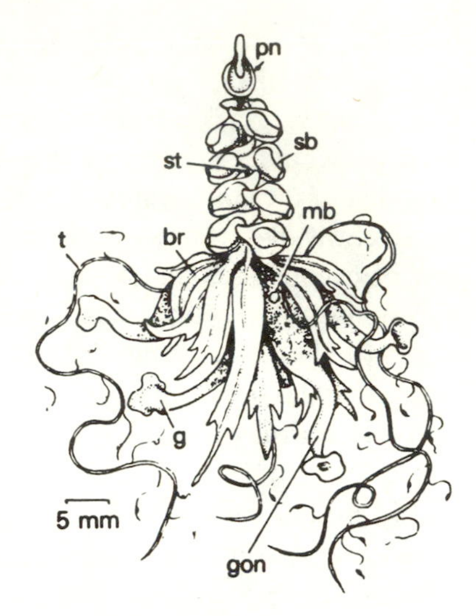

Fig. 50. The organization of a siphonophore. The air bladder at the top (pn) is a modified medusa, as are the swimming bells (sb) which are attached to a communal gut (st). The protective bracts (br) are highly modified medusae (mb is a medusa bud). The reproductive polyps (gon) are mixed with feeding polyps (g), which have large tentacles (t). (From Fretter and Graham, *A Functional Anatomy of Invertebrates*, 1976, Copyright © by Academic Press.)

ity. In clonal forms this relation does not hold, and some show no signs of senescence at all; they continue to grow and propagate without loss of vigor, barring some environmental catastrophe.

From this brief discussion it is obvious that I am mainly concerned here with aclonal forms. This does not lessen the importance of the clonal existence which, as we have seen, has played a major role in evolution. Furthermore, the existence of so many clonal organisms around us at present must mean that their strategy is still significant in the ecology of today. Its meaning for development is also important, because clonal animals and plants are able to mass-propagate asexually and therefore regeneration and regulative development are especially common.

Size and internal complexity

Internal complexity is judged by the number of cell types found in an organism. The reason, the reader will remember, is to make it easy to compare organisms of widely different structures; it is a convenient and simple method, although not without shortcomings. The biggest problem is deciding how many cell types there are in any one plant or animal, and the larger the organism the more acute this problem becomes. It is obvious, for instance, that a filament of a blue-green alga (more properly called a cyanobacterium) is made up of normal vegetative cells, spores (akinites), and heterocysts, but how is one to make an accurate estimate of cell differentiation within a mammal? Each tissue may seem relatively homogeneous, but it is in fact a composite of numerous cell types. The liver has a number of different kinds of cells, as does the kidney or any other internal organ. Consider the nervous system, including the brain: besides the cells associated with the neurons, such as Schwann cells and glial cells, there are certainly many different kinds of neurons with diverging structural and chemical properties. The problem of counting them with any kind of certainty is extremely difficult—perhaps impossible—and there is the further difficulty of determining what is the equivalent of the cell differentiation found in blue-green algae or in any other simple organism.

We are saved by the fact that for our purposes we need only relative figures, and these can be easily provided for various groups of organisms. A number of authors have tried to do so, and there is reasonable agreement among them. In Table 1 the rough estimates gleaned from the collective wisdom of these authors is summarized. I suspect that as one goes down the table, the larger the organisms, the greater the underestimation in the diversity of cell types. However, there is no problem in establishing their relative positions, their sequence. One other point should be made about this list. Among the lower plants it is important to break down the categories into smaller taxonomic groups. For instance, if we group all fungi, we would include simple molds such as *Mucor* and large mushrooms, which differ considerably in their number of cell types. Or, among the algae, a group such as the Volvocales or the Chlorococcales is clearly smaller and less complex (i.e., fewer cell types) than the brown algae, which include huge kelp. Therefore the taxonomic subdivisions used in Table 1 are set by the number of cell types so that any one category will include a narrow range of different cell types. The result is that in the upper end of the complexity scale the taxonomic groups become more inclusive. In the case of vertebrates, it is difficult to imagine that there is any great difference in the number of cell types for fish, amphibians, reptiles, birds, and mammals. What differences there are will be small enough to be drowned by the inaccuracy of our estimates, and therefore they are lumped together. Similar lumping for the same reason has been done for arthropods, molluscs, and annelids. Among plants the difficulty also arises between angiosperms and gymnosperms, and again they have been put together.

We are now ready to ask the key question: What is the relation of the size of an organism to its cell differentiation or internal complexity? In a previous study I considered the maximum size achieved for each cat-

TABLE 1

Number of cell types	*Organisms*	*Approximate weight in grams*
1	PPLO (mycoplasma)	10^{-14}
	Small cocci (bacteria)	$10^{-12}-10^{-13}$
	Protococcus (green alga)	10^{-8}
	Cyanobacteria	$10^{-5}-10^{-6}$
~2	Spore-forming bacteria	$10^{-11}-10^{-2}$
	Large cyanobacteria	10^{-4}
4	Molds (chytrids, mucor)	$10^{-7}-10^{-5}$
	Volvox	10^{-5}
	Ulva	5
7	Mushrooms	$10^{-5}-10^{5}$
	Kelp	10^{5}
~11	Sponge, Cnidaria	$10^{-4}-10^{5}$
~30	Angiosperm (Lemna)	$10^{-3}-10^{-4}$
	Giant sequoia	10^{7}
~55	Fairy flies	10^{-5}
	Giant squid	$10^{5}-10^{6}$
>120	Fish, frogs	10^{-2}
	Blue whale	10^{8}

egory of number of cell types; here I am examining the entire range of organism size, from the smallest to the largest, for any one level of cellular complexity. The result is shown in Figure 51.

One cell type. Let us begin with the organisms that have one cell type only. At the smallest end of the scale the most obvious examples come from bacteria that do not form spores. Their weight is estimated by assuming a specific gravity of one and measuring their volume. Small spherical coccus forms will weigh between 10^{-12} and 10^{-13} grams. Even further down the scale one has mycoplasmas which achieve a size of 10^{-14} grams. However, they are such highly specialized parasites that it may be unfair to include them, for they may be able to manage with little bulk because they make use of certain essentials from their

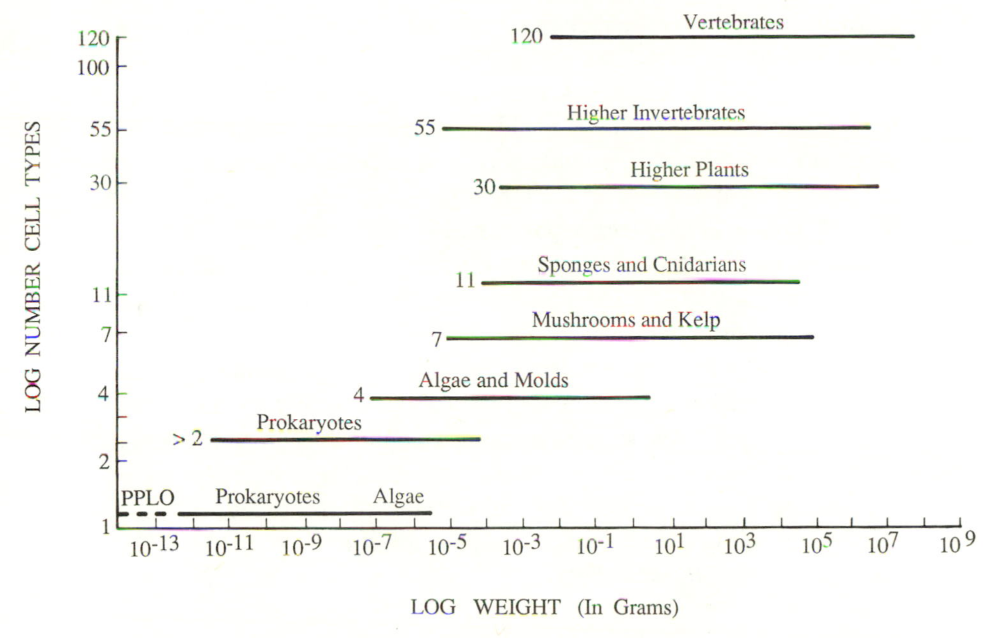

Fig. 51. A graph showing the size ranges (by weight) of groups of organisms containing different numbers of cell types.

hosts. For this reason they are put on a dashed line in Figure 51. The extreme case of such a trend is found in viruses, which lack the normal metabolic machinery of the cells and depend utterly on their host for energy and for the key molecules necessary for their own replication. All that is left is their nucleic acid (DNA or RNA, depending on the virus) and some protein, which serves various functions—for instance, it acts as a protective coat or container of the nucleic acid and helps in its transmission to new host cells. Because these are such a small part of a normal organism, they are not included in Figure 51.

Excluding multinucleate forms, the largest ''one-celled'' organisms are undoubtedly the ciliate protozoa, but they are excluded for obvious reasons. One *Paramecium* or a single individual of any ciliate possesses a huge macronucleus which contains the genome repeated many times, often two hundred or more. Therefore the ciliates must be considered organisms that have become large not by the unification of many single cells, each with a single genome, to form a multicellular organism, but by having many genomes in one giant nucleus within one giant cell. Ciliates are not comparable to the single-celled organisms we are considering; they are in a special class by themselves.

There are, however, eukaryotic organisms with only one cell type. Examples may be found among the algae. For instance, *Protococcus* is a simple green alga that grows on the bark of trees (Fig. 52). (It is often more prevalent on the windward side of a tree trunk, showing a telltale green streak.) Again using volume as a means of estimating weight, it is in the order of magnitude of 10^{-8} grams. *Protococcus* is asexual, for if it did have male and female gametes, it would immediately be put in a more complex category. Undoubtedly it is a reduced form descended from larger sexual ancestors. Therefore, some secondarily derived asexual algae fit in this one cell-type class.

If we look for the largest prokaryotes, then *Spirillum* would be a good example among bacteria with a weight of about 5×10^{-11} grams. Even larger single-cell types may be found among the blue-green algae (cyanobacteria) where simple filaments and small geometric figures of identical cells produce primitive multicellular colonies (Fig. 52). It is difficult to estimate the size of the colonies, but they should be within the range of 10^{-5} to 10^{-6} grams.

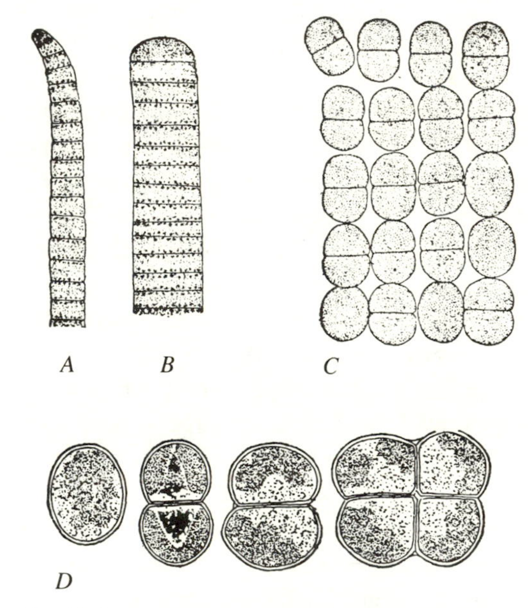

Fig. 52. Various organisms consisting of only one cell type. Three are cyanobacteria, *A* and *B* are different species of *Oscillatoria*, and *C* is *Merismopedia*. *D* is the green alga, *Protococcus*. (From Smith 1955, Copyright © by McGraw-Hill.)

Two or three cell types. The smallest organisms with two or three cell types would be spore-forming bacteria. A spore is a truly differentiated cell type and an example of a division of the labor. The vegetative cells do the multiplying and growing; the spores resist adverse conditions and will preserve the cell contents for the moment when favorable feeding and growing conditions return. In a fluctuating environment the advantages of such a resistant stage is obvious; this is a major evolutionary step. The smallest spore-forming bacteria weigh between 10^{-11} and 10^{-12} grams.

At the other end of the scale we find two kinds of large prokaryotes that possess two or three cell types. One is to be found among the innumerable examples of blue-green algae. Many species, which form long filaments, have periodic cells differing from their neighbors in that their internal cytoplasm appears clear (Fig. 53). These so-called heterocysts play a key role in the nitrogen metabolism of the filament and also serve as resistant spores. Blue-greens also have sporelike cells called akinetes. Filaments of many different species of blue-greens have both heterocysts and akinetes, and one might estimate the largest filaments to be about 10^{-4} grams. The other large prokaryote in this category are the myxobacteria (Fig. 53). Here the individual cells are true bacteria, but they feed in swarms of ever-increasing size. Ultimately the swarm produces a multicellular fruiting body that consists, in the larger forms, of a nonliving stalk made mainly of congealed slime and, depending on the species, a mass of spores or cysts at the end of the stalk. There is some suggestion, but as yet no firm evidence, that some cells specialize in the secreting of the stalk slime and die in the process, while the other cells go on to form spores or cysts. These fruiting bodies may reach a height of 1 millimeter and their estimated weight would be roughly 10^{-6} grams.

Four cell types. In this category we will include some simple algae and some fungi in which there are male and female gametes, and at least two different kinds of somatic cells. For instance, in the green alga *Volvox*, besides egg and sperm there are the ordinary vegetative cells and certain large cells (gonidia) in the colony that become the asexual colonies of the next generation; they alone are capable of cell division (Fig. 25). *Volvox* is a relatively small organism; its weight is roughly 10^{-6} grams. At the other end of the scale for algae, consider *Ulva*, the sea lettuce (Fig. 54). Besides its gametes, it has two other cell types: the

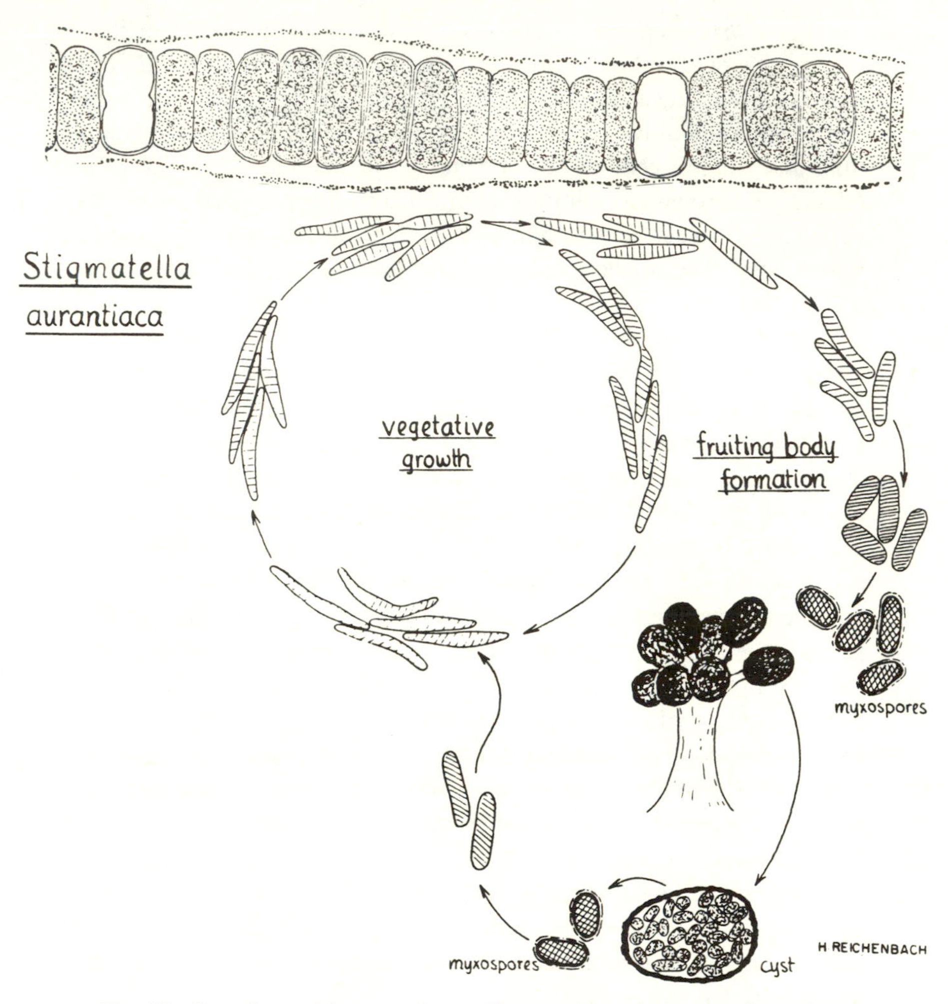

Fig. 53. Organisms with two or three cell types. *Above* is the cyanobacterium *Nodularia* showing vegetative cells, the thick walled spores, or akinetes, and the clear heterocysts which specialize in nitrogen fixation. *Below* is the life cycle of the myxobacterium *Stigmatella* which has both vegetative (feeding) cells and spores. (Above from Smith, *Cryptogamic Botany*, Vol. 1, 1955, Copyright © by McGraw-Hill; below, drawing by H. Reichenbach.)

cuboidal cells of the main leaf (thallus) of the plant and the filamentous cells which make up the holdfast. The size of a mature thallus of *Ulva* may reach that of a human hand, but it is only two cell layers thick, and its estimated weight is approximately 5 grams.

The filamentous fungi that have two distinct nonreproductive cell

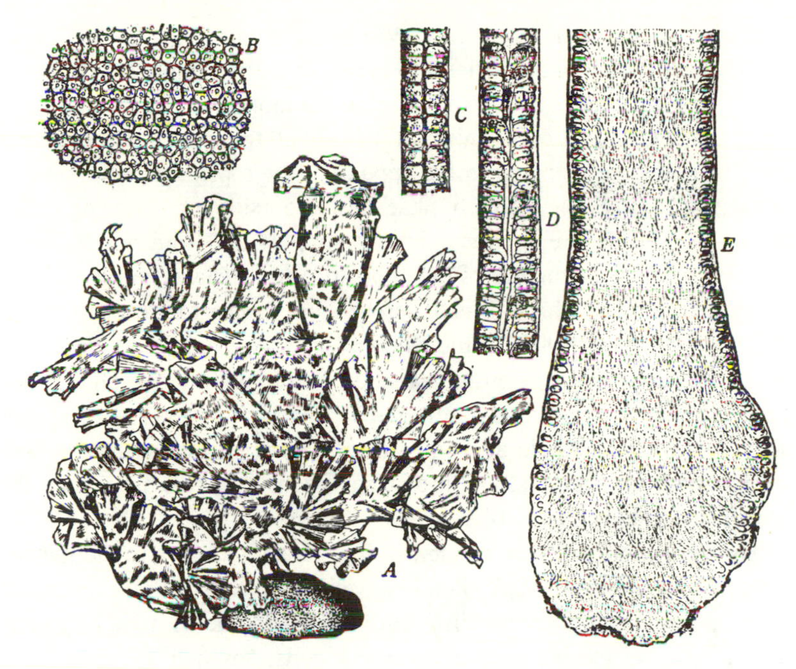

Fig. 54. The sea lettuce, *Ulva*, an organism with four cell types: the vegetative cells (*B* and *C*) and the supporting, filamentous cells found in the holdfast (*D* and *E*). The asexual zoospores and the sexual gametes are not shown. (From Brown, *The Plant Kingdom*, 1935, after Thuret, Copyright © by Silver, Burdett & Ginn.)

types are harder to weigh with any accuracy, because the cells are interconnections of filaments (hyphae) that grow continuously into a mat (mycelium); it is impossible to see where one individual begins and another ends because all are connected. Ultimately the protoplasm that has accumulated in the vegetative, or feeding, hyphae streams to one point and produces a very simple fruiting body in the smaller molds, such as *Mucor* (Fig. 28) or, as we shall see presently, it produces a large mushroom in more complex forms. The fruiting body is the accumulated protoplasm of one feeding area of the vegetative hyphae, and therefore its weight is the best approximation for our purposes. In *Mucor* and other simple molds the weight is about 10^{-5} grams. But some of the water molds, which also produce spores and vegetative hyphae, besides gametes, are even smaller, such as *Blastocladiella* or some of the chytrids that are in a weight class of 10^{-6} to 10^{-7} grams. It is difficult to know what might be the largest of the simple molds, but it certainly would be a small fraction of the weight of *Ulva*.

Approximately seven cell types. We are now entering a level of complexity where it becomes difficult to give a definitive figure for the number of cell types. If we look to some of the more complex basidiomycetes that form mushrooms, we find that not all the hyphae which make up the thallus are identical; there appears to be a primitive division of labor (Fig. 55). Some of the hyphae on the outside are covering cells, and others inside are involved in the support and transport of substances. Also, the structure of the spore-bearing surface (as on a gill in many species of mushroom) is different from that of the rest of the cap and from the stipe that holds the whole structure upright. But the differences are subtle and it is difficult to classify the different hyphae in the various regions into clear-cut cell types. Nevertheless, seven cell types seem to be a reasonable approximation. The smallest mushroom will weigh about 10^{-5} grams, and the largest giant puff ball may reach the quite remarkable size of 10^5 grams.

The very same problem of identifying the cells occurs in the marine brown algae, which include giant kelp, seaweeds that can reach massive proportions. Again there are different kinds of filaments in the stipe and the blades, but except for the "trumpet cells," which are thought to conduct nutrients great distances in these very long plants, it is hard to distinguish the cell types (Fig. 55). There are relatively few records of the weight of these algae, although in some species of *Macrocystis* found off the California coast the heaviest recorded weight is 136 kilograms, or roughly 10^5 grams, which is comparable to a giant puff ball.

Approximately nine to twelve cell types. We now turn momentarily to animals and look specifically at sponges and cnidarians, which seem to have approximately the same number of cell types. Unlike large fungi and algae, they do not have filaments but cells that are more compactly formed; nevertheless, it is not always easy to distinguish cell types. For instance, besides gametes, sponges also have covering cells and collar cells that line their flagellated chambers, and they have a host of different amoeba-like cells (amoebocytes) that are difficult to distinguish (Fig. 56). They contain granules of different colors, and the functions of some are easily identified; for instance, one type is responsible for the laying down of the spicules. In the case of cnidarians, there is a sharper delineation of cell types such as clear muscle cells, nerve cells, and stinging cells. The distinctions become more difficult when one examines the different cells that line the gut; they are not identical, and some are more directly involved than others in secreting digestive enzymes (Fig. 56).

The smallest sponges and cnidarians could be estimated to be roughly 10^{-4} grams. We are to an increasing degree dealing with organisms where we must consider the minimum size of an adult, not an embryonic or juvenile stage. We define adult simply as that stage or size at which the individual is capable of producing gametes, capable of starting a subsequent generation. As for the upper size limit, there are very large sponges, but it is sometimes difficult to know to what degree they are an amalgamation of individuals that have fused together. The largest aclonal cnidarian is certainly the giant species of jellyfish of the genus *Cyanea*, which reaches a diameter of over 2 meters, but there is a difficulty. A large portion of the body is inert jelly, so to have an accurate measure of ''living'' size is a problem. For both sponges and jellyfish, one might estimate the maximum size to be between 10^4 and 10^5 grams (10–100 kilograms).

Approximately thirty cell types. Higher plants, that is, gymnosperms and angiosperms, fall into this category. Because gymnosperms such as pine trees and other conifers have slightly more simplified wood (xylem) and bast (phloem) tissues than angiosperms, the latter probably have a few more cell types than the former; but the difference is slight so that it is sensible to lump them into the same category. One difficulty with estimating the number of cell types is that there are many different structures in plants, such as the different parts of the flowers in angiosperms; yet if we look at histological sections, we see no obvious differences in the cell structure. Clearly differences must exist, and therefore it is probable that thirty is a great underestimate of the total number of cell types in higher plants. But, as I explained earlier, there is no need to resolve this thorny problem; all we need to establish is that higher plants have more cell types than sponges and cnidarians, and this is unequivocally the case.

In the size range of higher plants, the minimum is in the order of 10^{-4} grams. This range applies, for example, to duckweed (*Lemna*), which floats as small leaves on pond and marsh water, but it has fewer cell types than larger angiosperms. It has very simple roots that protrude below the surface, and a rudimentary stem. Giant sequoias are the largest gymnosperm, and some species of eucalyptus in Australia, which are angiosperms, become almost as large. The difficulty with estimating

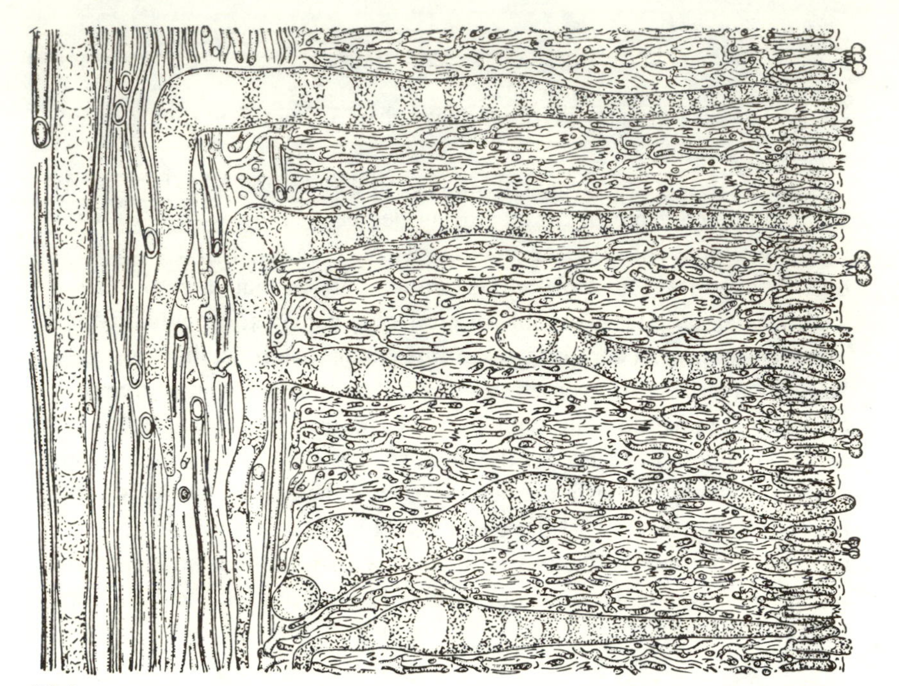

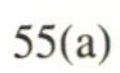

55(a)

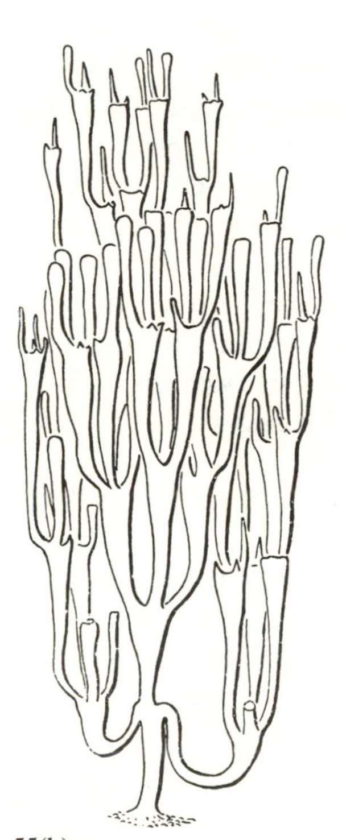

55(b)

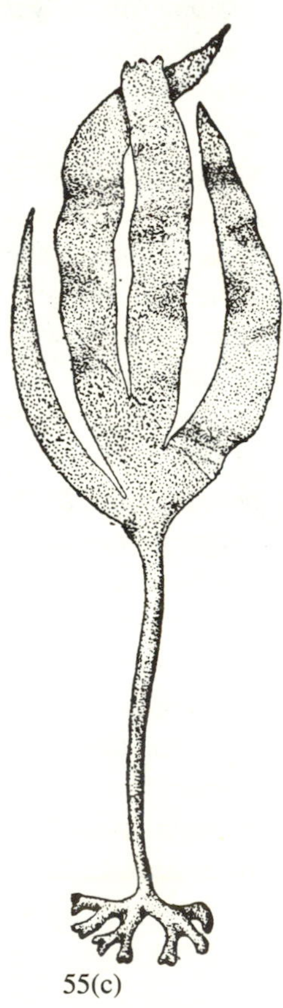

55(c)

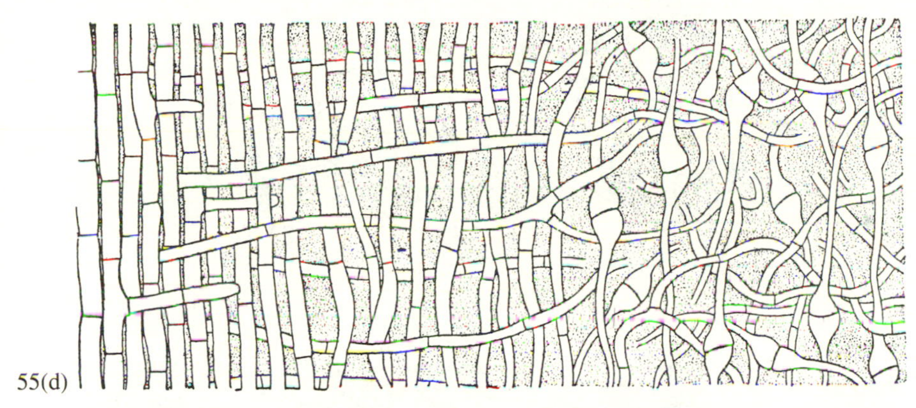

Fig. 55. Organisms with approximately seven cell types. (a) A section showing the variety of kinds of hyphal elements (and the basidiospores) in a species of fungus closely related to *Clavaria*. (b) The appearance of the whole fruiting body of the same species (from Corner, *Clavaria and Allied Genera*, 1950, Copyright © by Oxford University Press.) (c) A large brown alga or kelp (*Laminaria*) and (d) a section through its stalk showing different cell types. (Smith, *Cryptogamic Botany*, Vol. 1, 1955, Copyright © by McGraw-Hill.)

the weight of such trees, which are about 100 meters tall, is that, like the giant jellyfish, a large part of their bulk is not alive—in this case, of course, it is dead wood. It is, however, possible to estimate the amount of live cells in a tree, and I am indebted to Dr. John Grace of the University of Edinburgh for showing me how to do this. Approximately 90 percent of the leaves of a tree are made up of live cells, as are 9 percent of the trunk, stems, and roots. One makes many assumptions in such calculations, including how these ratios might vary with size, and a rough estimate for a 100-meter sequoia is 10^7 grams of live cells.

Approximately fifty-five cell types. We now wade into far more difficult territory—that of higher invertebrates, which include insects and other arthropods, annelid worms, molluscs, and echinoderms. The main point is that it is impossible to make fine distinctions between all these different groups of organisms with respect to their cell-type number. They clearly have more cell types than higher plants, and after consulting various authorities on invertebrate anatomy, I find that fifty-five types seems to be a reasonable estimate and can be used for our working purposes.

Within this huge spectrum of different kinds of organisms, insects are the smallest. For instance, small wasps called fairy flies, which are 0.5 millimeter long, weigh in the order of magnitude of 10^{-5} grams.

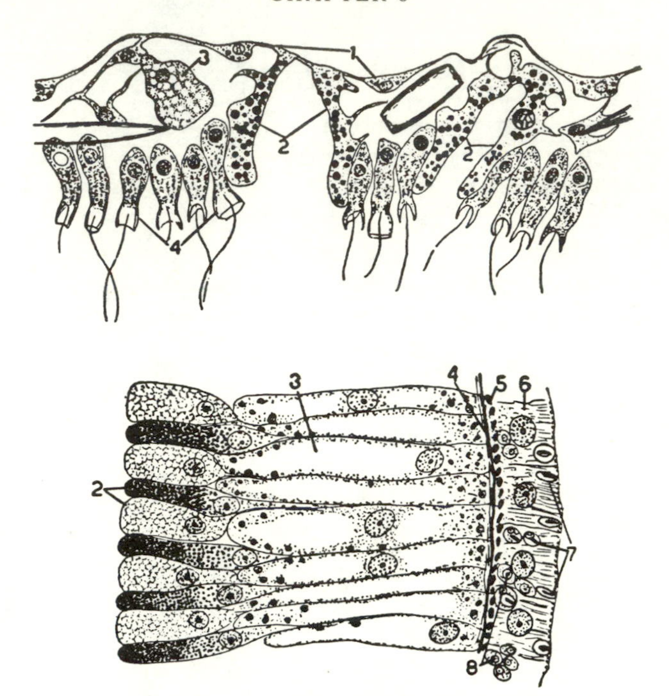

Fig. 56. Detailed sections through parts of organisms with approximately eleven cell types. *Above*, a cross-section through a freshwater sponge showing four cell types. (1, covering cells; 2, pore cells; 3, an amoebocyte; 4, a collar cell.) *Below*, a section through the mouth region of the cnidarian *Hydra* showing six cell types (2, two types of gland cells; 3, epithelial cells; 4, muscle bases of epithelial cells; 5, muscle bases of epidermal cells; 6, epidermal cells; 7, stinging cells; 8, interstitial or undifferentiated cells.) (From Hyman 1940, Copyright © by McGraw-Hill.)

No one has done the histology of fairy flies, at least that I know of, and one assumes that they have the same number of cell types as large wasps and bees. It would be interesting to know if this assumption is true. It is not so in some species of very small rotifers, where the gut cells are reduced and the gut becomes a large mass of continuous protoplasm, rather like the insides of a protozoan.

The largest complex invertebrate is undoubtedly the giant squid whose weight lies between 10^5 and 10^6 grams. It is the same order of magnitude as an elephant.

More than 120 cell types. Again because fine distinctions are impossible, we shall lump all vertebrates together. No doubt fish have fewer cell types than mammals, but establishing this fact firmly might take a lifetime of sterile investigation. The reason for thinking 120 is an underestimate of the number of cell types is that it is based primarily on morphology, and there is every indication that some cells may be identical morphologically yet quite different in their physiological or biochemical processes. For instance, I am told by an expert in the field that it is possible to recognize ten to twenty morphologically distinct neurons in the central nervous system of a mammal, but that the number could possibly rise to between fifty and one hundred if it included neurons with different neurotransmitter and receptor properties that show no morphological differences. Therefore if we were to include nonvisible chemodifferentiation of cells, perhaps vertebrates might have up to five hundred or more cell types. Again, the actual amount is not crucial, but rather the fact that vertebrates have significantly more cell types than invertebrates. Obviously this cannot always have been true, and perhaps there is a closeness of cell-type numbers in primitive vertebrates such as lamprey eels (cyclostomes) and advanced invertebrates such as squid (cephalopods). But we shall ignore the overlap and consider the two groups as major separate entities.

The smallest vertebrate is probably some minute species of fish, although there are also exceedingly small frogs. 10^{-2} grams is a reasonable rough value for the minimum size of vertebrates. The largest is clearly the blue whale, which reaches a weight of 100 tons or 10^8 grams.

All the information on the relation between the number of cell types and the minima and maxima of organism size has been plotted on Figure 51, which we must now examine in more detail. It would be interesting to know the mean size for each cell-type category, but this is impossible. It is important to remember, however, that within a taxonomic group, or even within a large collection of taxonomic groups, there is a characteristic frequency distribution of species of different sizes, as we saw in Chapter 5 (Figs. 42 to 46). There is a maximum number of species somewhere near the lower end of the size scale and then it falls off, abruptly for even smaller species and gradually with larger ones. Presumably one could draw such a frequency-distribution curve on each of the horizontal lines on Figure 51, but one still would not know exactly where the frequency distribution maximum lay for each cell-type level. Were this possible one might assume that the maximum peak of the frequency distribution represents the optimum size for a particular level of complexity. Perhaps during the course of evolution, when a new level of complexity was reached, the first organisms were of a size that had particular advantages.

A glance at Figure 51 makes it obvious that whether one considers minimum or maximum size, or the hypothetical optimum, there is a direct correlation between complexity, as measured by the number of cell types, and size. In a taxonomic group of organisms of a particular size range, the larger the size, the greater the number of cell types. Because there is such inaccuracy in estimating this number, no meaning can be attached to the slope of the line.

Cell size and internal complexity

There is another aspect of size and complexity that I would like to discuss briefly. It is the question of the relation between cell size and complexity in multicellular organisms. We have already examined the peculiar case of ciliate protozoa, where the cell size becomes very large and the cell structures highly complex and differentiated, but the nucleus is compound and repeats the entire genetic content or genome of the organism many times. Here I want to examine those cases that consist of standard multicellular organisms made up of cells, each with a single genome.

Many people, including D'Arcy Thompson in his *On Growth and Form* (1942), have made the point that there is a consistency of cell size that is independent of the size of the organism, and the size of a liver cell or a neuron of a mouse is approximately the same as that of an elephant. The difference in the overall size of the two organisms is entirely reflected in cell number.

In general, there is a very close correlation between the size of the genome and cell size, a point that was already suggested in the previous discussions of ciliate protozoa. In the next section I will look into the relation of genome complexity to organism complexity; here the emphasis is solely on cell size. In a well-known series of experiments, G. Fankhauser (1955) studied cell size of newts of different ploidies, that is, newts that could be experimentally induced to develop even though they had half the normal complement of chromosomes in the body cells (haploid, 1N), as compared to the normal diploid (2N), or three sets of chromosomes (triploid, 3N), four (tetraploid, 4N), and even five (pentaploid, 5N). The remarkable aspect of these newts is that they all have the same body size, and therefore the haploids will have numerous small cells, while pentaploids will have fewer but larger cells. This is so despite the fact that, as Fankhauser showed, all the internal organs are of normal proportions. This is strikingly illustrated in the kidney tubules, where the proportions are identical, yet the cell sizes differ greatly (Fig. 57). It should also be mentioned that some plants can also be produced in different ploidies, and in their case both the cell sizes and the whole plant size increase together with the increase in ploidy—a further variation on the theme.

There are, then, exceptions to the rule that we first enunciated: that organism size is independent of cell size and is solely reflected in cell number. However, they do not in any way undermine the general principle that cell size is not an important influence in organism size. At least this is true of Fankhauser's newts, where the final size and all the detailed internal shape is the same regardless of differences in cell size. We can therefore conclude that cell size has little direct bearing on the complexity of the organism; the only possible relevance is that it reflects the genome size, or the number of genes, and we will now see how gene number affects organism complexity.

Genome size and internal complexity

A great deal has been written about the amount of DNA per cell in organisms of different complexities. The interest stems from the fact that there is an enormous variation in the amount of DNA in different groups of organisms. The general consensus is that there is no direct correlation between complexity and the genome size of an organism. This is well illustrated in Figure 58, which makes it quite obvious that the overlap of DNA per cell for different groups of organisms is extremely large. Furthermore, if one looks at a particular group, such as algae or amphibians, the range in genome size is quite extraordinary.

One of the main interests in the amount of DNA has centered around the question of how much is involved in directing the development of the organism and other active processes such as metabolism, and how much is redundant or passive and carried along as so much baggage. It has been pointed out that only the former DNA is relevant to the question of complexity, and the inactive DNA obscures how much active DNA is present. If one could measure the total messenger RNA—the product of the active DNA responsible for coding the proteins that are synthesized in the cytoplasm of the cell—then one would expect to find something that would correlate with complexity.

As far as our main theme is concerned, we are left with the strong

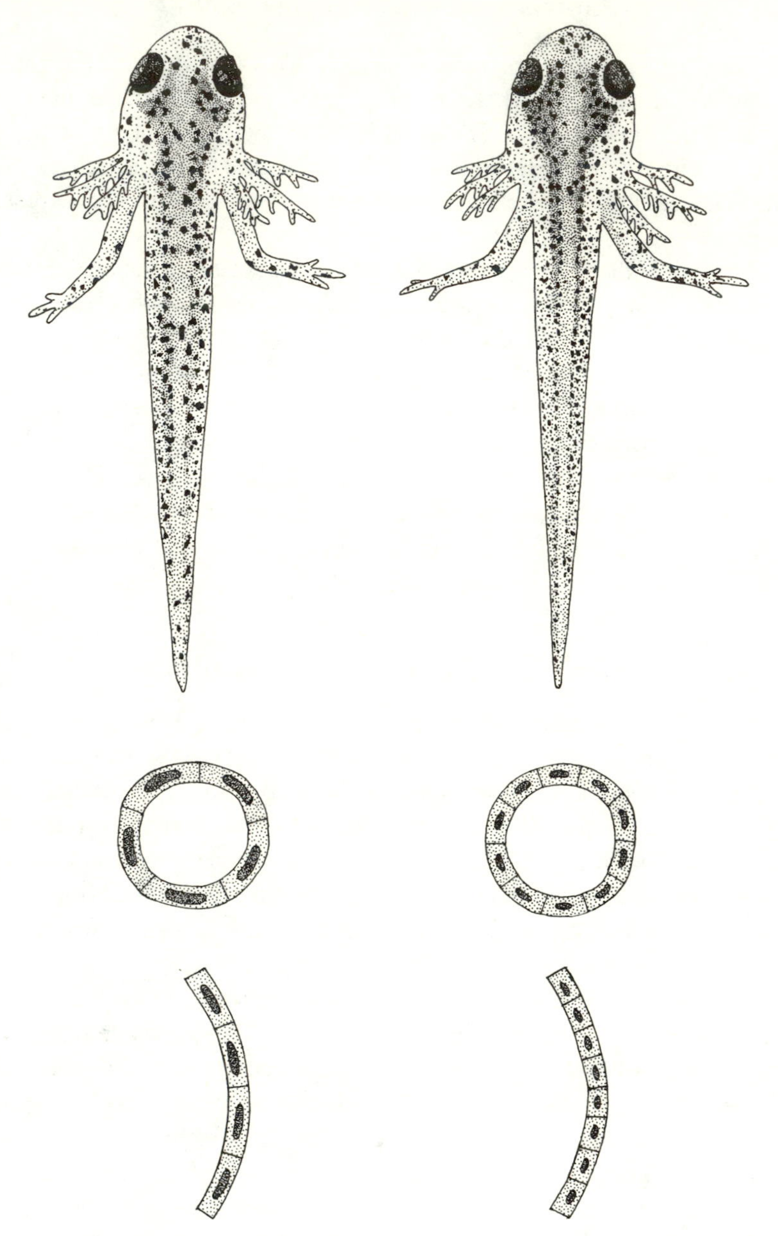

Fig. 57. Two larvae of the salamander *Triturus viridescens*. The pentaploid larva with five sets of chromosomes (*left*) is the same size as the diploid larva with two sets (*right*). As can be seen from the circular cross-section of a kidney tubule in the middle and a covering layer of cells over the lens, the tissues also are the same size despite the striking differences in cell size. (From photographs kindly supplied by Dr. G. Fankhauser.) (Bonner 1980, Copyright © by Princeton University Press. Drawing by Margaret La Farge.)

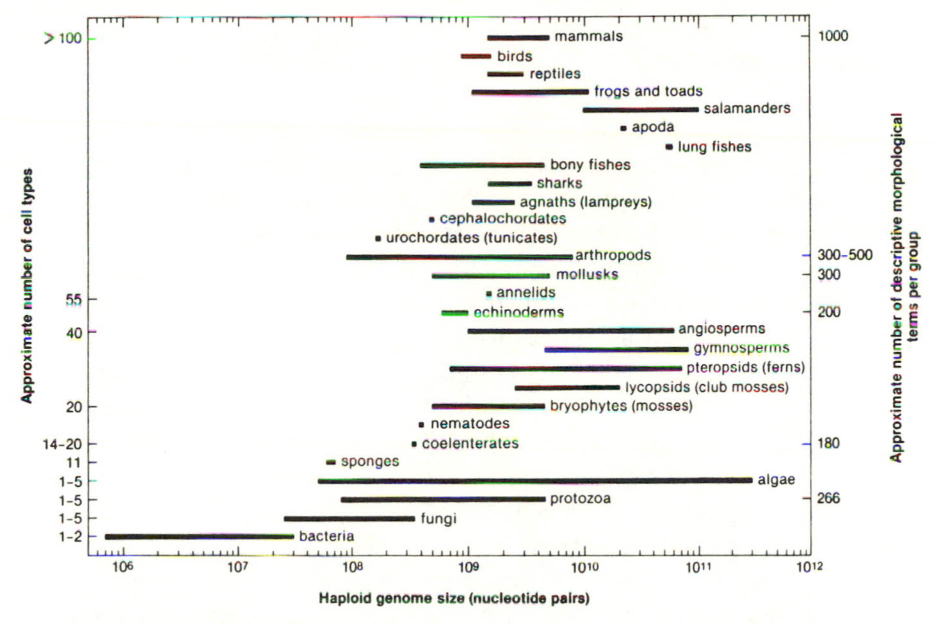

Fig. 58. The genome size of groups of organisms of different degrees of complexity. Note that there is little correlation between the two. (From Raff and Kaufman 1983, Copyright © by Macmillan.)

conclusion that it is the size of the whole organism that correlates best with the internal division of labor or complexity. However, on any one level of complexity (as measured by the number of cell types), the range of possible organism sizes is enormous.

How is a large, complex organism built?

Earlier I asked the relatively narrow question of how, mechanically, a large multicellular organism is produced, and I stressed that, with the multiplication of cells during development, there must be some sort of adhesive or skeletal method that keeps the cells together in the right shape, ultimately making them capable of supporting themselves when subjected to the force of gravity. I admitted at the time that this was an artificial procedure and that it was impossible to separate differentiation (complexity) from the mechanics of size increase because the appearance of adhesive sites on certain cells, and the selective secretion of strategically placed portions of skeletal material, all involve differentiation. Size increase and the increase in complexity occur simultaneously during development. Now let us consider development primarily from the point of view of the arising of the division of labor, or internal complexity, which includes the differentiation connected with the support system.

As I have discussed elsewhere in some detail, the easiest way to think of development is as a process which involves three main constructive elements. The first is growth, which is generally reflected in cell multiplication in multicellular organisms. The second is morphogenetic movements, which can only occur where there are mobile cells and, therefore, is essentially absent in higher plants with their rigid cell walls, but prevalent in animals with their amoeboid embryonic cells. These formative movements are responsible for the shaping of the embryo; they are involved in such basic processes as gastrulation and neurulation, major steps in forming the fundamental body plan of the animal-to-be. The third constructive process is differentiation, which is, as has been repeatedly mentioned, the basis of the division of labor or complexity within an organism. Differentiation can occur in rigid plants that seem only to have patterns of growth as a means of achieving shape, and it can occur in animals in ways that sometimes seem to be quite dissociated from growth.

For me the best way to think of the problem of differentiation is to think of the consistent pattern that develops each generation. This so-called "pattern formation" means that a substance is *produced* at the right *time* and in the right *place*. First, let us begin with the production of substances. If cells become different, that is, divide the labor and form different cell types, they have special, characteristic, individual chemical properties. This simply says that muscle cells, nerve cells, and cells of the lens of the eye, for instance, are quite different in certain of their chemical constituents; somehow the different constituents were stimulated to be produced in particular cells. But different cell types do not arise at the same time; some appear sooner than others. Therefore, there must be a mechanism that tells a cell when to produce key substances, as well as what substances. Finally, any specific group of cells must not only appear at the right time, but at the right place. In many ways the process of "localization" is especially important, for this is the ultimate key to producing a pattern. Lens protein must be produced only in the lens itself, and not in the liver, or the muscle.

If development, and especially differentiation, is thought of in this way, then the developmental biologist seeks mechanisms whereby synthesis, the timing of the synthesis, and the placing or localization of the synthesis are rigidly controlled and consistent from generation to generation. The approaches to the problem of these mechanisms are two-fold. One is to look for overall developmental signaling systems that play crucial roles in turning differentiation on, or in some instances turning it off. The signals might be hormones or "inductors" that stimulate, or a variety of inhibitors that counter or prevent the stimulation. These substances might diffuse in the form of a gradient and in this way have different effects, depending on what part of the gradient hits a certain region. It is known that it takes more than these negative or positive stimulating substances to produce an effect. Cells have receptor proteins, substances that combine with the signal molecule; and the combined receptor-signal molecule in turn instructs the cell to respond in a particular way (or not to respond, depending on the signal). These receptor molecules can in turn be distributed in a pattern by localization, so that a general, unpatterned signal can produce a complex pattern. In other words, there is an interplay between signals—which may or may not be in a pattern—and receptors which may be evenly or unevenly distributed in some organized way. The interplay of such a stimulus-response system, with all its potential powers of spatial distribution, is the macro-organization system of development, and it is the feature of development most carefully studied by developmental biologists.

But there is another important aspect of development which is, so to speak, the other end of the problem. It is generally agreed that the development of any animal or plant is encoded in the genes. There is also some information in the egg cytoplasm, but that also comes from genes; it was merely synthesized previously. What is needed, therefore, is some way of understanding how genes, which are responsible for the synthesis of specific proteins, can give out protein signals in such a way that they can control all the complexities of synthesis of other kinds of substances. And the timing of all the syntheses in the embryo must also be controlled. In other words, these syntheses, as we said initially, must be coordinated so that they can cause a particular substance to be produced at the right time and in the right place. The initial part of the process, namely the steps immediately following the activation of the genes and the mechanism of activation, are the subjects that draw the prime interest and attention of the developmental geneticist. What is needed today is some way of bringing these two important areas of inquiry closer together so that the gap which presently separates genes from the ultimate pattern can be closed. What I will do here is discuss that gap in very general terms in order to give at least a conceptual framework of how it might be bridged.

The gap between developmental genetics and development is presently a concern of many, as G. S. Stent has eloquently and repeatedly pointed out. Recently, R. S. Edgar suggested to me that with the tools of modern molecular genetics, we are now in a position to try to understand what instructions are sent out from the genome that determine the developmental program of the organism. It is clearly not just a matter of identifying structural genes, that is, genes which are responsible for producing a specific protein, but understanding how regulatory genes affect the activity of other genes. It is important, for instance, to find out how one gene can have many effects (pleiotropy), and how such effects can be both on development and on some fundamental process in the metabolism of the cells (the so-called housekeeping functions of cellular activity). We may learn something from the study of multigene families, where a gene is repeated, often with slight differences in the sequence of their component nucleotides. There are some obvious reasons for having many of the same genes; for instance, it is a convenient way of making more of some protein that is needed in large quantities. But Edgar also suggested that it might be a way of initiating differentiation, for the small differences in the structure of the sibling genes could be used to provide different proteins in different cell types.

Finally, he made the point that it might be exceedingly useful to approach the problem from an evolutionary point of view. The use of modern molecular genetic techniques should make it possible to determine which genes are common to all organisms, and which differ between major taxonomic groups. For instance, both animal and plant cells have similar metabolic pathways of oxygen consumption and energy processing, yet they differ in the presence or absence of a cell wall and in many related aspects of their pattern of growth. If one could compare the genes that differ in various taxonomic groups, either widely separated or closely associated, one might be able to identify those genes primarily involved in a particular kind of developmental program.

Clearly the exciting new developmental genetic work being done with the fruit fly (*Drosophila*) and the nematode (*Caenorhabdites*) and other organisms is an attempt to bridge the gap. But it is still no more than a first step, albeit a most desirable one. Here we are concerned with the entire gap between genetics and development, and the work so far involves only a portion of it. The difficulty is that we are wandering blindly in an unknown territory and can make only wild guesses as to what we will find. It is perhaps better not to invent a whole complicated hypothetical scheme, but to try to understand the elements of development and imagine how they could be manipulated by genetic signals.

To begin, we must remember that genes can be responsible for the synthesis of many different kinds of proteins. Some of those genes will be structural, that is, they will produce proteins directly, and those proteins will be components of the cell. They may have a wide variety of functions: for instance, they may be involved in the skeleton of the cell, as are molecules such as actin or myosin, or they may be enzymes, protein catalysts that are responsible for carrying out an enormous variety of chemical reactions.

Equally important are the genes that appear to be involved in the control of gene activity itself. These are regulatory genes. From the early work of F. Jacob and J. Monod on gene activity in the bacterium *E. coli*, it is known that whether a gene is turned on or off, that is, whether it makes messenger RNA or not, is often under active control. In some genes there is a special protein called a repressor which prevents the genes from transcribing messenger RNA; and for transcription to start, the repressor is blocked by yet another regulatory molecule.

The next step in complexity is the organization of various genes so that they can either fire off in concert or perhaps follow one another in some sort of sequence. In other words, there is an organization of the activity of the genes within the genome that provides the first step for the organization of the process of development. The general principle is that the genes not only regulate their own activity, but some genes can also control the activity of others. There is among eukaryotic organisms a hierarchy of gene activity that undoubtedly plays a crucial role in development. Among the first to make specific hypotheses explaining the regulation of gene expression in eukaryotes were R. J. Britten and E. H. Davidson (1969) in an important paper. Their basic concept is one of a network of genes in which the activity of some have a specific stimulating or inhibiting effect on others. This is again a stimulus-response network as discussed above, but in this case we are not concerned with hormones or induction substances and their receptor molecules; we are considering processes that occur within the genome. However, we must remember that all the cells within a developing multicellular organism have the same genome, and somehow these gene activities have to be coordinated among nuclei. It would be of no use to have this complex activity between genes occurring in one cell and not in others.

For communication between nuclei, clearly messenger substances are necessary, and so we come back to hormones and inductors. That hormones can induce the expression of specific genes in insects has been known for a long time. In the fruit fly (*Drosophila*) and other flies, the salivary glands of the larvae have giant chromosomes in which each gene is repeated many times (Fig. 59). When the molting hormone (ecdysone) either appears normally before the molt or is injected by experiment, certain specific bands puff, as can be seen in Figure 59. By using radioactively labeled nucleotides, which are the building blocks of messenger RNA, it is possible to show where RNA synthesis (that is, transcription) takes place, because only those regions where the label is permanently incorporated in a large messenger RNA molecule will show radioactivity. In Figure 59 all the puffs, which are at specific gene bands, show active transcriptions, and there is none elsewhere at any of the other gene sites. The ecdysone has acted at specific sites, and as a result a few specific messenger RNA's have been synthesized. These in turn are synthesized or translated into specific proteins in the cytoplasm. (There is recent evidence that the first set of puffs induces a second set on some other genes, whose products are responsible for turning off the first set, thereby preventing the initial stimulus from being permanently turned on.) Unfortunately, we do not know what those proteins are, nor how they lead to molting of the larva; if we did, we might have another large insight into the mechanics of complex development.

We have shown that there are ways in which some genes can affect the activities of others, and that there are signal mechanisms so the same event can occur in all the nuclei in the body. This leads us to two further points of fundamental importance: one is the extent of the stimulus-response network which has its origin in gene activity, and the other is the even greater problem of localization.

As far as the networks are concerned, one must first remember that many of the protein products of the genes are enzymes, and that they catalyse many different chemical reactions. If one considers the contents of a cell, one sees that besides proteins there are an enormous amount of other substances: macromolecules such as carbohydrates, fats, and a whole variety of small molecules, many of which are hormones, vitamins, and substances involved in the energy-metabolism machinery of the cell. Furthermore, this complex of protein enzymes, substrates, and products has a life of its own—that is, there are all sorts of regulatory mechanisms known to exist at a level entirely beyond the gene. In enzyme reactions the activity of the enzyme can be affected by the amount of substrate or the amount of product, or often by the

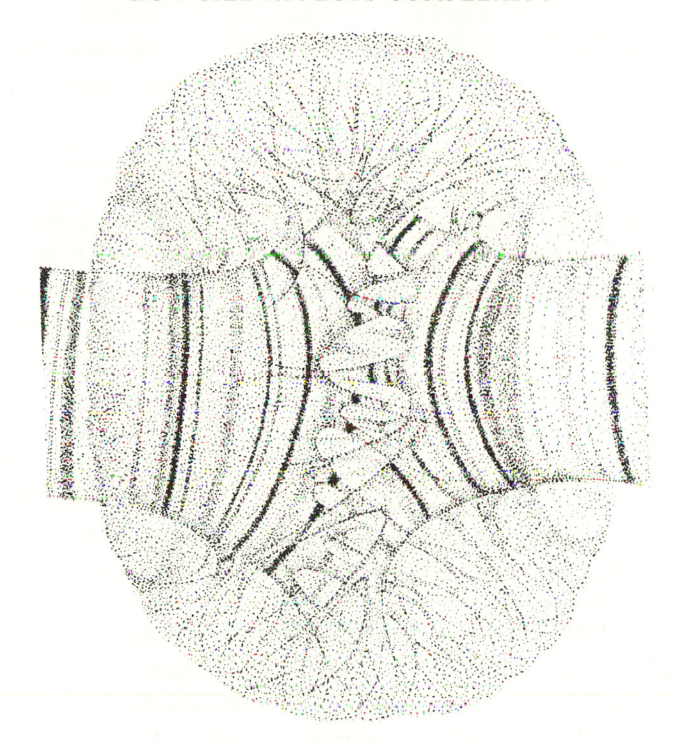

Fig. 59. A puff on a portion of a giant chromosome of a fly. This is a region of active gene transcription in a specific part of the chromosome. (From Beerman and Clever 1964, Copyright © by *Scientific American*.)

amount of product of some subsequent reaction in a sort of cascade of chemical steps. The way in which a product or any other substance can affect an enzyme is quite remarkable. The substance might be in excess, for example, and some of it may combine chemically with the enzyme, whereby it changes the shape of the enzyme (allostery) and makes it less capable, or perhaps totally incapable, of activity; or, in other instances, it might make the enzyme more reactive. Shortly I will return to the significance of allostery, but first we must continue with the description of the chemical network.

Some of the gene products beget yet other substances in subsequent enzymatically controlled steps, and some of these substances will interact. It is often the case that if two different products interact by chemical combination, an enzyme is needed for the reaction to take place. Certainly many enzyme reactions occur in sequences where the substrate of one reaction is the product of a previous one. And remember that each of these products has the ability to interact with other substances in the cell. Often those reactions are specific, and again the result of one might affect one or more other specific reactions in turn. Since there are very large numbers of enzymes and a still larger number of other substances in even the simplest cells, there is the possibility of a vast network of interconnecting and controlling or regulating reactions. It is as though relatively few genes have generated a complex giant of chemical interactions that, once launched, can do remarkable things on their own. Later we will call this a ''gene net'' and we will discuss it at a higher hierarchical level. Since the gene effects can be multiplied and ramified in this way, it is easy to see that it would be very difficult to predict their course if all one knew were the nucleotide sequences of the various active genes. In the genetic control of cellular synthetic activity, the genes have produced a monster far bigger and more complex than is suggested by their nature and composition. But it is not a monster out of control. Quite to the contrary, every secondary and tertiary reaction, and all those that follow, and all those that involve the cross-reactions of different gene products are potentially under gene control. A slight alteration in a gene may have insignificant or far-reaching consequences, for the possibility of its effects being multiple is very large (pleiotropy). If those effects are far reaching and if they do produce a monster of the wrong sort, it will not survive. The only place where the gigantic network can be effectively changed is at the level of the gene, and such gene changes may result in network changes that are selectively advantageous; this is the way evolution occurs. It is also the reason why evolutionary change cannot be Lamarckian; one cannot just change some of the secondary steps in the network and then pass the change on to the next generation. The sole changes that are heritable are gene changes; they alone can be passed on from one generation to the next.

Let us return to allostery and the matter of the difference between genes and their products. In allostery the shape of a protein enzyme can be modified by chemical combination with another substance, in this way affecting its chemical reactivity. When a substrate combines with an enzyme, the enzyme may become deformed so that any further chemical activity is impossible or altered. Clearly then the shape of proteins is of unique importance for their activity, and this is true not only of enzymes, but also of other proteins such as the contractile proteins in muscle.

Genes only indirectly specify this three-dimensional or tertiary structure of proteins. They specify a sequence of amino acids, and the fact that the strings of amino acids will coil and wrap themselves into a tight ball of a particular shape is, in a sense, a fortuitous consequence of the amino acid sequence (Fig. 60). Particular amino acids along the chain will combine with complementary ones elsewhere on the chain, and the result will be a folding of the chain into a characteristic shape. The important point is that the chain's tertiary shape is a consequence of its sequence of nucleotides. So the linear genes of the DNA give rise to all sorts of other effects, of which the three-dimensional tertiary structure of proteins is one. The form of the folded protein has new properties that would not exist if the amino acids simply stretched out in a straight line. These properties involve specific kinds of chemical binding that permit enzyme activity (and allostery), contractility, and other activities characteristic of proteins. Furthermore, these derived, three-dimensional features are ultimately responsible for the gigantic network of chemical reactions that are the fabric of development.

The next question is even more interesting. How can these gene products, which produce a complex of interacting chemical pathways within a single cell, act in a coordinated fashion among many cells of identical gene composition in a developing multicellular organism? Genes do not act in the same way everywhere, but there is a pattern to the synthesis of some substances. The transcription of a gene or one of the many chemical reactions following protein synthesis varies in different regions of the embryo. In other words, the network of pathways from the gene to the cell cytoplasm (and the temporal sequence of those pathways) is not the same for all the cells; differences appear at different times in different cells or groups of cells. This is yet another order of complexity which rises above the complexity within cells. It would appear that the pattern of chemical reactions during development becomes increasingly remote from the primordial coding of proteins by genes. There are so many steps in between—and these chemical steps seem to take on such remarkable new properties as they become increasingly large and complex—that it is hard to remember that the ultimate control of all this complexity lies in the genes.

Since evolution has occurred by natural selection, it is essential there should be a system of inheritance that can go through a single cell stage, and therefore the genetic code must be the master controlling agent for all the complexities of development and the ultimate adult structure. This is such a fundamental point that we will return to it again and

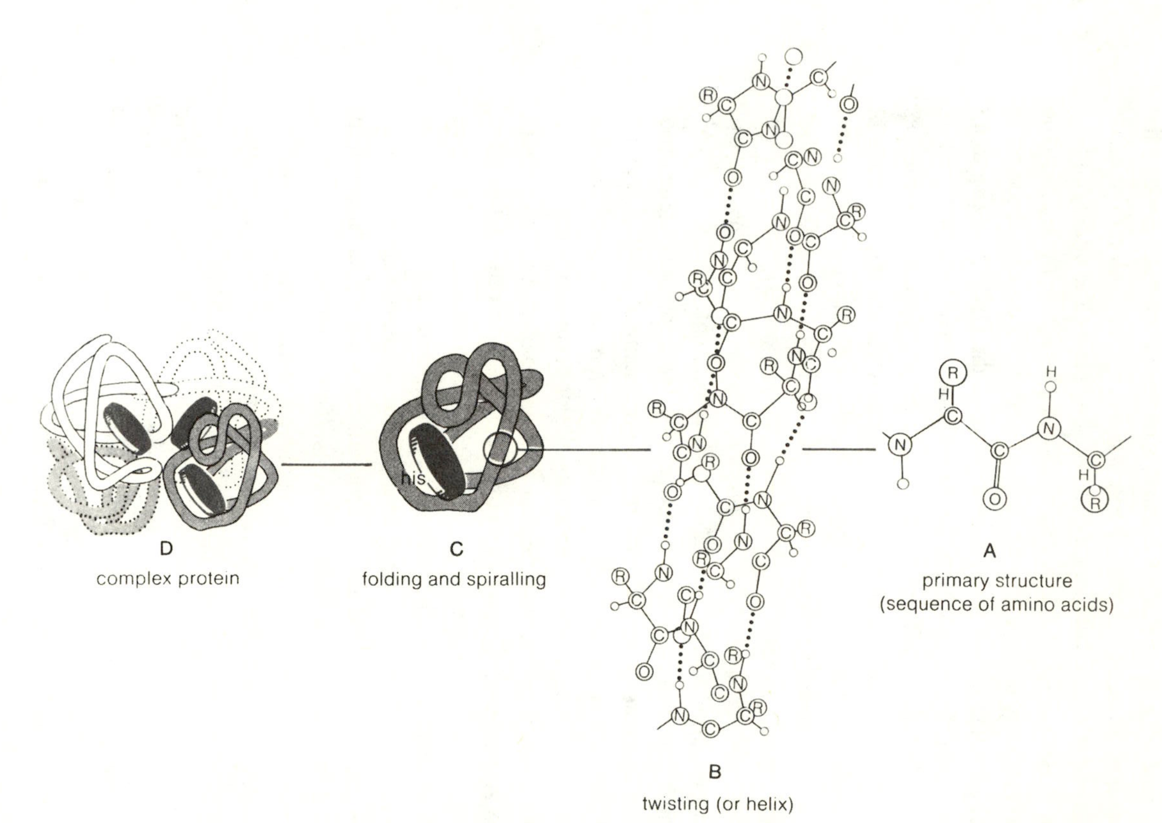

Fig. 60. The sequence of amino acids is responsible for the ultimate three-dimensional structure of a protein. Interactions between atoms projecting from the amino acids cause the string of amino acids to twist, fold, and spiral to form a protein with unique properties. This protein can be bound with other proteins to form a complex protein structure such as hemoglobin shown here. (From Steyaert, *Life and Patterns of Order*, 1971, Copyright © by McGraw-Hill.)

examine it in more detail. The fact that genes ultimately control all the hierarchical levels of complexity we find in development is not a trivial generalization, but a profound one: natural selection would have shaped it no other way. Although hierarchical levels of complexity may have new properties at different levels, those emergent properties are not mystical, but direct consequences of increasing complexity. They must all ultimately be explained by the characteristics of the genes, just as the explanation of the new shape and qualities of a folded protein molecule are to be found in the sequence of its amino acids, which in turn is specified by the nucleotide sequences in the gene. The difficulty is mentally to encompass the extraordinary ramifications that are derived from a simple sequence of nucleotides.

We still have not come to grips with the problem of pattern formation and localization, and how these might be controlled by genes. It is an impossibly large problem, and I will only seek to identify the rudiments. These rudiments fall into two general categories. One is the signal response system and the other is the spacing of both the signal substances and their receptors. The subject of chemical signals and their receptors has already been briefly discussed. Here I want only to show how these systems can be related to genes. Signals and receptors are chemical substances, so their very existence is the result of gene activity. Since some signals and all receptor molecules are proteins, they usually are the direct product of the transcribed and translated gene, or they could be proteins that have become modified by another enzyme. In any case, since they are gene-specified proteins they have the capacity of extraordinary specificity. This means that a receptor may respond only to a particular signal and in this way shows powerful properties of discrimination. In many instances the signal substance is some small molecule such as indole acetic acid, the growth hormone in plants, or slightly larger ones such as the steroid sex hormones in mammals. Such small molecules are also derived from direct gene products through a whole series of enzyme-controlled reactions. They provide a nonspecific stimulus, as one would expect from small nonprotein molecules. This is in contrast to their receptor molecules, which are proteins with specificity and will react only with one particular signal substance. Therefore one direct connection between genes and signal-response systems is that the molecules themselves are molded under orders from the genes, and these orders permit a great variety of chemical structure and therefore specificity.

An orderly development, however, does not automatically follow from making a set of perfectly fashioned signal and receptor molecules; something very basic is still lacking. For one, these molecules cannot be produced all at one moment in development, but must only appear at the appropriate time, as in the case of the molting hormone ecdysone in the insect larva, which appears in quantity only at certain stages in the growth of the larva. Therefore order, in part, comes from a timing sequence. In recent studies on early amphibian development, M. Kirschner, J. Newport, and J. Gerhart (1985) have proposed that there are at least three separate mechanisms. One of these controls the cell cycle, and there is evidence that this control is associated with a cytoplasmic factor, a protein that is released at regular intervals. Another independent timing device controls a radical change that occurs to all the cells in the midblastula stage, and this one seems to be triggered by achieving a critical nuclear-cytoplasmic ratio as a result of successive cleavages. The third clock is responsible for the onset of gastrulation, and although less is known about its mechanism, there is evidence that it involves the translation of maternal messenger RNA's into proteins.

It is quite possible that there is always more than one timing mechanism within developing cells. One should not forget that all organisms, or at least an extraordinarily wide variety of those that have been tested, from unicellular forms to higher plants and higher animals, are capable of keeping time with great accuracy. Unfortunately, we have virtually no understanding of the molecular basis of these timing mechanisms, although there have been some excellent mathematical models for circadian clocks. In principle, timing is an easy problem to solve, for one can set up a hypothetical series of sequential chemical reactions that cycle in some way, and this regular cycle can then be monitored by some adjoining chemical system. The difficulty has been in dissecting out and identifying such a clock from a cell.

Besides the temporal pattern of the release of gene products, or secondarily derived substances, there is also the equally vital matter of the spatial pattern. Here we have had a great assist from mathematicians, for they have made it possible to see, in a wide variety of models, how a pattern can emerge even in cases where substances may originally be evenly distributed. By far the most important influence has come from the reaction-diffusion models and their relatives, the ideas of which have their roots in the early work of N. Rashevsky and were made more generally acceptable by the pioneering work of A. M. Turing (1952).

To illustrate, in reaction-diffusion models assume that one or more substances, as for instance an activator or an inhibitor, are capable of diffusing at different rates within the organism. The very fact that one is called an activator and the other an inhibitor shows that the substances are not inert, but react, and for this reason they are known as reaction-diffusion models. By making various assumptions about the kinds of molecules—that is, their chemical activities and diffusion properties—one can make all sorts of predictions of how these substances will be distributed when they reach a steady state. It is also theoretically possible to greatly increase the variety of patterns by increasing the number of these key substances, or morphogens, as they are sometimes known. In some organisms, such as the cnidarian *Hydra*, or in the cellular slime molds, some of the morphogens have been identified, and this hunt for key signaling molecules in development is very active today, just as the hunt for the chemical identity of mammalian and plant hormones has been a matter of intense interest and success for many years. What we end with is a solution to the problem in principle with a few of the key molecules identified, but with never enough chemical detail to be able to have a totally satisfying analysis of pattern.

The last point I wish to make about localization of substances in the developing organism is that not all cases fit the kind of *regulative development* modeled and described above, where every part of the embryo has the potential to develop into a complete new whole. There is an equally important kind of localization that comes under the general terms of *mosaic development*, or differentiation by cell lineage. In this case there are certain substances, certain gene products, either direct or indirect, that are confined to particular cells in the embryo and they do not diffuse to other cells; they are partitioned off in their cellular prisons. A classic example is in the development of *Styela*, an ascidian, originally described by E. G. Conklin at the turn of the century. It has the advantage that one can see certain areas of the cytoplasm in the fertilized egg that have different natural colors, and as cleavage proceeds in early embryogenesis, these different areas of pigmented cytoplasm become contained or packaged in different cells (Fig. 61). By following the sequences of the cleavages, as Conklin did in wonderful detail, one can trace exactly which cells will form muscle, nerve cells, and so forth; one can follow the lineages of each cell in the completed larva. The same kind of phenomenon is found in many other invertebrates, such as molluscs, nematode worms, and numerous others, while

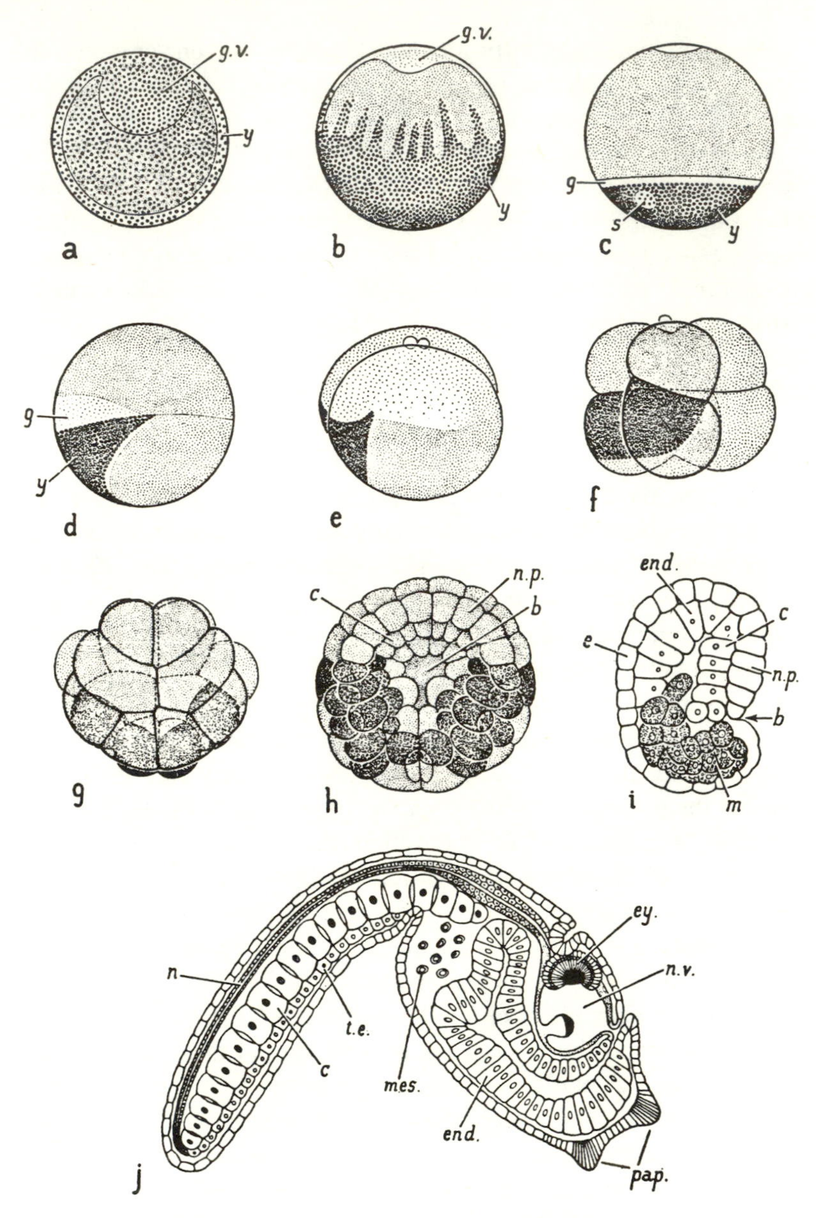

Fig. 61. The development of the ascidian *Styela*. (a) Before fertilization showing an even distribution of pigment. (b) Immediately after fertilization the pigment moves into specific regions (c, d), which correspond to particular tissues of the larva. The fate of these pigments can be followed right through development to the larva (j); it is a classic example of mosaic development first described by E. G. Conklin. (From Waddington, *Principles of Embryology*, 1956, Copyright © by Unwin Hyman.)

cnidarians, echinoderms, and all vertebrates largely have regulative development. However, there are two interesting qualifications to this general statement.

One is that many of the examples of extreme mosaic development have a period of regulative development at some other stage in their life cycle. For instance, in *Styela* before fertilization the pigments are evenly distributed in the egg, and then they fall into definite regions in a matter of minutes after fertilization (Fig. 61). This rapid distribution of key substances is really the primary period of localization, and all further localization is due to the movement of the cells carrying these different substances. It has been suggested that this initial distribution of substances in the egg might be due to diffusion; if that is the case, it would invoke the same physical force that we have suggested plays a significant role in localization in regulative development. Indeed, as A. M. Dalcq (1938) showed some years ago, an early egg, before pigment redistribution, is regulative, and any portion of that egg, provided it contains the nucleus, will produce a normal, miniature embryo. Furthermore, the adult ascidian shows remarkable powers of regulation, and one can regenerate an entire individual from just a few cells.

Another mixture of mosaic and regulative development is illustrated particularly well in *Hydra*. Most of *Hydra*'s development is regulative except for the differentiation of certain cell types. These are the derivatives of the small so-called ''i-cells'' which become either nerve cells or stinging cells (nematocytes), depending on how many divisions they have undergone, as C. N. David and R. D. Campbell have shown (Fig. 62). This is a very rigid cell lineage, and it is controlled by the need to replace either nematocytes or nerve cells. There is a feedback mechanism that turns the division sequence on or off. Notice also that some cells are always kept as i-cells; they are a permanent source of the other cells. This cell lineage system works side by side with all the other cells of *Hydra*, which are totally regulative.

Later we will ask the evolutionary question of why we have such extraordinarily diverse methods of localization, including mosaic and regulative development; here I want merely to describe that variety, and show in broad strokes how it might be related to the genetic information in the nucleus and its immediate products. To put the matter as succinctly as possible, there are timing mechanisms which are responsible for the release of substances. We do not understand the nature of those clocks, although we can well imagine that either through cyclic chemi-

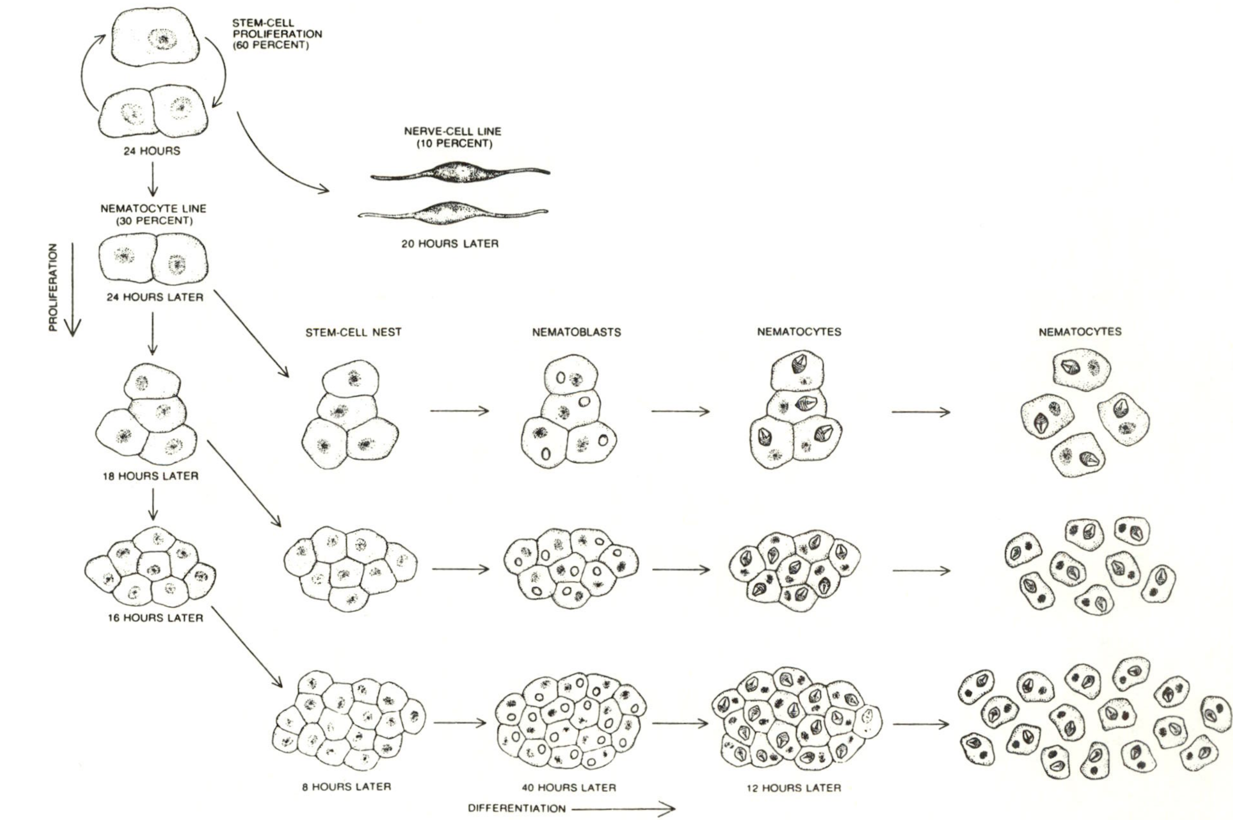

Fig. 62. Cell differentiation of nerve and stinging cells (nematocytes) in *Hydra*. As C. N. David and R. D. Campbell showed, stem cells (i-cells) produce both cell types, but the cell cycles are controlled in such a way that the two cell types appear in a fixed proportion. Each day about 60 percent of the stem cells reproduce their own kind. Another 10 percent differentiate into nerve cells, and 30 percent end up as nematocytes. In the first stages, lasting two to four days, the stem cells destined to produce nematocytes simply proliferate, typically forming nests of 4, 8, or 16 cells. Following the last division the cells in the nests start differentiating. After 8 hours the nematoblast stage is reached, showing the vacuole of the developing nematocysts. After 40 hours the nematoblasts have differentiated into nematocytes containing a mature nematocyst capsule. The nests finally break up, enabling nematocytes to migrate individually to the tentacles. (From Gierer 1974, Copyright © by *Scientific American*.)

cal pathways or a sequence of synthetic reactions, a particular product might be produced at a particular time. Since substances react and diffuse, they can be localized in space. This is easiest in those cases where the substances in question can transcend cell boundaries and give gradients in cell masses. But many of the signals, and certainly all the receptors, are large molecules, and their distribution is dependent on the activities of individual cells to which they are largely confined. It could be that different cells have different receptors because they had previously received a signal from a small diffusing molecule to synthesize that particular receptor. Another possibility is that the macromolecular signal-response systems are passed down by cell lineage; they can be present in certain cells and not in others because of their line of descent from different ancestral cells, as in *Styela*. Or perhaps these cells can become localized with their special macromolecules because the cells are capable of movement and can travel to the appropriate places. Such movement must in turn be directed, and it often is by diffusion gradients of some small signal molecule. In other words, the entire processes of synthesis, timing of synthesis, and localization can depend on the interactions of the substances themselves. There are various restrictions, such as the size of the signal molecule or the size of its receptor, and these determine how the substances will interact with one another. That they do so is determined by their chemical structure, and even though that determination may be (but not necessarily) remote from the first gene-product protein, it is still totally determined by that gene product. Signals and responses can occur between gene products, descendants of gene products, other genes, and in all the combinations of a vast network of chemicals distributed in time and space within the developing organism. But those networks and all their complex interrelations are affected by genes, which pull all the strings of the marionettes, however long and branched those strings might be. This is the direct result of natural selection, for as I emphasized earlier, there has been a selection for a one-cell stage in the life cycle which permits sexuality and gene recombination to occur, and a selection for large adults which see to it that the genetic products of the sexual reproduction are protected and are in turn able to achieve success in reproduction in subsequent generations. For this reason the genes, which can be neatly packaged in the single cell stage of the cycle, must carry all the instructions necessary for development and for the rest of the life cycle. (This ignores the information, especially in the form of proteins, that messenger RNA's and other substances present in the egg cytoplasm, but these are also gene products from previous life cycles.)

Plasticity in Development

So far the argument has implied that every aspect of development, even in large, complex organisms, is rigidly controlled by genes. This has been a convenience to demonstrate a principle; now we can examine an interesting group of exceptions. In some organisms there are many events in development that are not genetically determined, or at least not completely so. Either the gene products have a stochastic variability to them, or they can be pushed one way or the other by the environment. In both cases they are within a genetic frame, and sometimes it is easy to see what might be the selective advantage for the plasticity, but not always. I shall give a number of examples of this kind of developmental plasticity.

One of the most interesting examples has been discussed in some detail by M. J. Katz (1982, 1983). A number of workers have shown that in vertebrates more neurons are formed than are used in the final nervous system. After the complete system is fully developed, those nerve cells that have not been incorporated into it die. The evidence favors the idea that the cells form their synaptic connections with other cells, and as the whole nervous system begins to function, those synapses and those cells that are part of the functional circuits remain; the rest wither away. It is assumed that all the cells during their development send out sprouts in the form of axons and dendrites, and that in this way a vast number of interconnections are made, involving a considerable amount of chance as to which cells become connected. It is thought that if the genes had to specify each connection, which amounts to staggering numbers in a vertebrate, neural development would be an impossibly complex problem for even the largest genome. Therefore all that the genes specify is the differentiation of the neurons and their general initial placement and direction of growth (and some of these directions are likely to involve small molecule signals and various other gene products far removed from the genome itself). There also is specificity in the adhesions between different types of neurons involving particular molecular combinations on the cell surfaces, all of which would be gene-specified for the particular cell types. How the neurons connect with one another would then be largely chance, and some form of function would establish the final, permanent pattern.

One reason for the interest in this kind of development, as Katz so rightly emphasizes, is that it illustrates how remote a final structure may be from the genetic information found in the genome. In this case chance can, within the proper, genetically determined setting, produce a variety of different detailed patterns. One wonders if, among human beings, this might not be the basis of the extraordinarily wide differences in our intelligences, or our musical talents, or our skills with languages. From our point of view the important message is that by having such a stochastic step in pattern formation, the genes have produced a clever way to generate a complexity far greater than they can manage by more conventional developmental processes.

How this mechanism arose during the course of evolution presents an especially intriguing problem. Presumably in the simplest nervous systems in invertebrates there was some advantage to having the nerve cell actively function, in that way making certain pathways in the nerve able to pass messages with great fluency. This is presumed to be an element of memory. Once this principle is established, which it clearly is in invertebrates, then by increasing the number of neurons and circuits, function can remain the key organizing principle. Such being the case, it makes no difference if the cells grow fibers that connect at random or in a fashion preordained by the genes; in either case they would be firmly established by active function. Since unnecessary genetic programming is expensive, it is easy to see how it might have been supplanted by chance. And this early switch in simple nervous systems means that ultimately animals could be produced with enormous brains, far too large to be genetically programmed in every detail.

Another kind of plasticity is illustrated particularly well in plants and has been comprehensively reviewed by A. D. Bradshaw (1965). There are many instances where clearly one particular aspect of the morphology of a plant is not only variable, but its variability is directly influenced by the environment. To give some examples, a Mediterranean grass, when grown under conditions of low and high fertility, will show an enormous increase in the number of seeds in richer soil, but the seed size varies little. Other plants, grown in different soils and at different altitudes, will show great variation in height and the number of shoots, leaves, and flowers but little variation in such characters as leaf shape and flower structure. Therefore, not only do the plants vary with the environment, but only specific parts of the plant vary; other parts are stable. Bradshaw concludes that the ability to show morphological plasticity is genetically determined (since only some features are plastic and others are not), and that it is the plasticity of the development of the plant that is selectively advantageous. This is especially evident in cases where the environmental changes may occur over periods of time less than a full generation, or over short distances in space. Then a plant can cope with these short periods or small distances far more effectively by phenotypic variation rather than by any kind of genetic change and selection, which take at least one generation to adapt and in most instances many more.

The extreme cases of environmental effects are found in shallow-water aquatic plants. Their leaves may be submerged, at the surface, or in the air, and the difference is important for the physiology of the leaf. In water buttercups (crowfoot), submerged leaves are like fine hairs, while the floating leaves are solid (Fig. 63), and in breeding experiments it has been shown that this plastic response to water is inherited and therefore genetically determined. Here the adaptive advantage of such plasticity is even more obvious. As I. I. Schmalhausen (1949) shows, this kind of effect also exists in animals, and an excellent example discovered recently is the effect of crowding on the shell shape of snails growing along the coast (P. Kemp and M. D. Bertness 1984).

If we relate these instances of plasticity, either involving chance or responses to the environment, to our previous discussion of development, it is clear that there are some events which occur beyond the direct control of genes, yet they always seem to be encouraged by the genes and invariably the available options are set in part by the genes. This is because in all instances they appear adaptive for one reason or another. The most significant point for our grand argument is that if one looks at animal behavior or leaf shape in the water buttercup (or growth forms in hydroids, bryozoans, or molluscs under conditions of different food abundance), it would be impossible to identify any genes which directly produce plastic behaviors. The phenotype has taken on properties that are not rigidly specified in the genome, in that way adding a new dimension to the complexities of development.

Social insects. As a final example I will describe the development of social insects. They add all kinds of plasticities onto development and at the same time provide a wonderful insight into development itself.

An ideal example is found among ants when, after the nuptual flight, the queen will seek a suitable nesting site. She sheds her wings and uses the degenerating flight muscles to provide energy to the small larvae

Fig. 63. The water crowfoot showing the great difference between the broad aerial leaves and the thin submerged leaves.
(Drawing by S. Ross-Craig, *Drawings of British Plants*, 1951.)

which emerge from the initial group of eggs she has laid. As soon as these workers pupate and molt into adults, they become helpers to the gigantic queen, their mother. If it is a harvester ant it will scurry out to find small seeds to bring into the nest, and assist the queen in feeding the new emerging larvae. Because now more food energy is entering the nest, the next generation is larger and eventually there is a whole range of different-sized workers (Fig. 64). Once these morphological differences are manifest, the workers begin to specialize in different duties: the smallest ones remain in the nursery; the medium-sized workers become the foragers; and the largest ones remain behind to defend

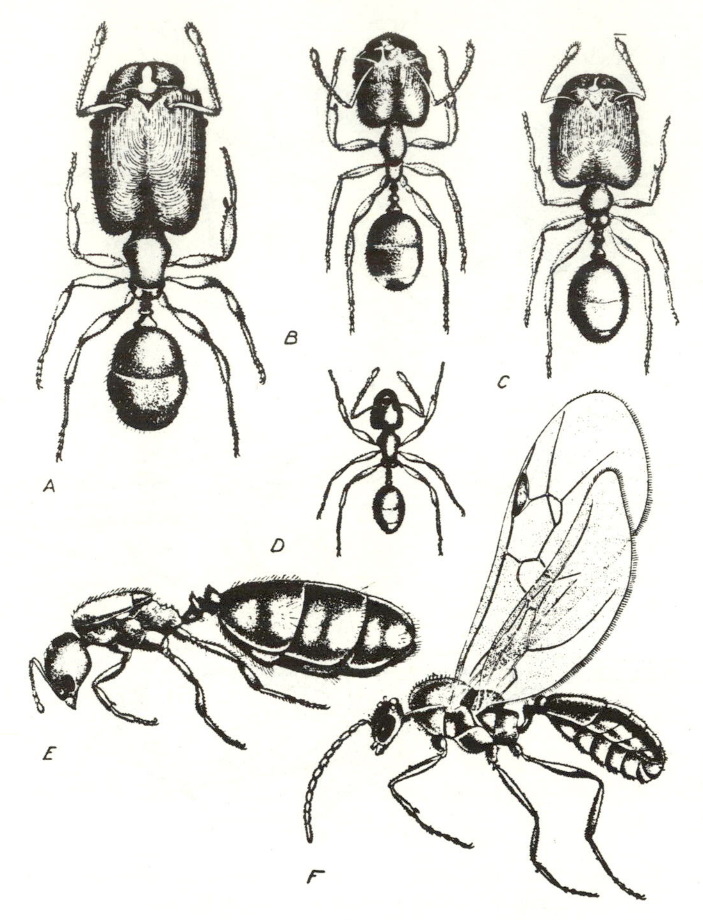

Fig. 64. The different castes or morphs of the ant *Pheidole instabilis*. *A–D*, different size sterile workers; *E*, a queen; *F*, a male. (From Grassé, *Traité de Zoologie*, Vol. X, 1957, after W. M. Wheeler.)

the nest: they are the soldiers. All these are skills originally performed by the queen when she was alone, but now with so many daughters to help, she concentrates on egg laying.

It has been established beyond any doubt for both ants and termites, which have been intensively studied, that the enormous differences in the size and the structure of the workers are entirely environmental; the genes for their development are the same from the smallest worker to the largest soldier, even though the jaw shape and other structural features of the soldier might appear quite different. The only individuals that are genetically different are the short-lived winged males; they have half the number of chromosomes (haploid) compared to the females (diploid). The latter include the queen and all the nonreproductive workers and soldiers.

Numerous factors govern the size (and therefore the shape, for the two are linked) of the workers, but all of them are environmental. In the example just given, clearly the richness of the diet has played the key role. It is also known that the role of chemical substances passed between individuals (pheromones) is a factor, shown especially clearly in the study of termites. These pheromones usually are inhibitors and, for instance, if there is an excess of soldiers in a termite colony, the existing ones will produce sufficient pheromone to inhibit any of the nymphs from molting into soldiers. If all, or a large portion, of the soldiers are removed, so will be the inhibitor, and as a result there will be many new soldiers appearing at the subsequent molt. In this way a ratio is maintained between the various classes of workers, and these ratios remain relatively constant regardless of the size of the colony. The pheromones provide a feedback mechanism which regulates the relative numbers of the different castes very much like the inductors and hormones that provide the correct proportion of parts during the development of any individual. They provide a kind of pattern formation within the colony, or gigantic family, as it might more appropriately be called. But the pattern is not a spatial one, for all the individuals are moving about; it is a pattern in the abstract sense.

A particularly interesting insight as to how pheromones affect morphology came from the work of M. Lüscher. He showed that the difference between workers and soldiers in one species of termite was due to the balance between two hormones within an individual: the juvenile hormone which stimulates growth rather than differentiation, and the molting hormone (ecdysone) which is more of a differentiation hormone. If the latter is high, a soldier will emerge, but if the juvenile hormone is relatively high a worker will emerge. The balance between these internal hormones is in turn governed by the external pheromones, so there are two levels of chemical signals, one inside and one outside the individual organisms. The pheromones treat the organisms as though they were cells, and the internal hormones do indeed act directly on the cells. I have already shown that ecdysone can act on the genome and cause puffing in the chromosomes, reflecting the transcription of specific genes. This leads one to the picture of a great web of chemical conversations taking place between organisms, within organisms, within cells, and as stated earlier, even within nuclei. Furthermore, substances from any one of these levels can reach across and converse with others at any of the other levels. So our stimulus-response mechanisms seem to have no boundaries: they go from societies to genes.

The relevance of this example of insect societies to my main argument is that it provides yet another illustration of how far removed the ultimate effects of genes can be from the genome itself. Not only are the adult forms of ants and termites plastic in the sense of the structure of the nervous system or the different forms of crowfoot leaves, but they also are made possible by genes. The ability to produce pheromones, and the ability to respond to them, is again something laid down in the genome of the fertilized egg. In this respect it is no different from any other developmental signal-response system. Even if the signal is not hormonal, but due to diet, or temperature, or light, the response system has to be specified by genes and their products. How else could it appear again at the next generation from a single-cell fertilized egg? For not only is a whole ant derived from one cell, but to a degree so is an ant colony. The queen comes from such a cell, and although the eggs she lays are fertilized by sperm from a male, all the workers are her progeny, and their genetic difference may be small and certainly insignificant to our argument: the huge colony comes from single cells. (It is interesting that this is not true of all insect societies, and many, such as honey bees and army ants, divide the workers between a new queen and the old one. In this sense such insect societies are like an individual organism, for they divide by binary fission.)

The Evolution of Internal Complexity

What we have shown so far is that development may be complex, and one manifestation of the complexity is that there are ways in which the ultimate gene products can show nongenetic variability, either of a stochastic nature or in response to environmental circumstances. Furthermore, this plasticity can extend to the morphology of separate individuals in the interesting and special case of insect societies. Now we must examine the evolution of complexity—how it progressed from primitive beginnings. Following this we will be in a position to compare the problem of evolutionary change in small, simple organisms with those of large, complex ones.

If one looks at the fossil record, or more correctly at our reconstruction of the sequences of organisms, one comes to the inescapable conclusion that the earliest forms were simple and small, and that as evolution progressed there appeared increasingly complex and larger plants and animals (Figs. 1, 5). We can also deduce from the discussion in the beginning of this chapter that there is a very rough correlation between size and complexity measured in the number of cell types (Fig. 51). The correlation is probably rough because although increased complexity is, within limits, required for size increase, the reverse is not true. There can be an increase in complexity without any increase in size, or even more commonly there can be a decrease in size without any change in complexity, thus obscuring what is probably a fundamental relation between the two.

One important point discussed earlier is a purely mechanical one: size increase requires some division of labor. It is dictated by mechanical considerations, as is true in any machine that must function as a unit with increased size. In living organisms it is a direct consequence of the fact that some properties go up as the cube of the linear dimensions, such as weight and the amount of tissue that is metabolized; and others, such as strength, gas exchange, and food assimilation, go up as the square of the linear dimensions. The result is an invention of new cell types that handle locomotion and the coordination of locomotion and all sorts of ways to cope with internal metabolism, such as gut systems and respiratory, circulatory, and excretory systems. The problems become most acute in animals because of their mobility, but the principle applies to stationary plants as well; they also are beset with weight-strength problems and metabolism-diffusion problems, as well as problems of transporting fluids against gravity.

Another element that plays a key role in the relation of size to complexity is natural selection. It is true that selection sees to it that the mechanical considerations described in the previous paragraph are followed, and that any deviations will be at a disadvantage, but selection is important in an even more direct way. There can be selection for size and there can be selection for complexity, and the two can be independent of each other. We have already discussed selection for size; here we will concentrate on complexity.

The simplest case to imagine is the transition from organisms of one cell type to those with two. In spore-forming bacteria or amoebae, the advantage of a second cell type in the form of a spore is obvious. These minute organisms are totally unprotected in their environment if they are free living and not parasites, and this means that they are constantly subjected to dramatic changes in their environment. It can be a switch from a thaw to a freeze, or from on excess of water to extreme desiccation. If they are nothing but one type of growing or vegetative cell, these extreme changes could easily wipe out an entire patch. Therefore the selection pressure to produce a resistant stage would be great, and any mutation that produced some more resistant cells would be preserved. The simplest arrangement (and the one most often found) is that in which all the cells go into the resistant state; they are triggered by, for instance, early signs of desiccation, or more often by the depletion of their source of food, and these signals of hard times ahead promote sporulation. Therefore in these instances the pattern of differentiation is temporal, for at one time all the cells are of one cell type, and later some or all are of the other.

The advantages of having two cell types appear together at the same time are quite different. In filamentous blue-green algae (cyanobacteria), the heterocysts, which appear at regular intervals within a chain of cells, represent a division of labor of metabolic functions (Fig. 53). Photosynthetic plants must have nitrogen in order to synthesize their proteins, nucleic acids, and other essential nitrogen-containing compounds. If they cannot obtain sufficient nitrogen compounds from their immediate surroundings they must grab (or ''fix'') atmospheric nitrogen gas. The difficulty is that the enzyme responsible for nitrogen fixation (nitrogenase) cannot operate in the presence of oxygen, a substance that is the direct product of photosynthesis. The problem has been solved in cyanobacteria with the isolation of all the nitrogen-fixing chemical apparatus into the thick-walled heterocysts, and the photosynthesis machinery in the surrounding vegetative cells; this is a perfect division of labor forced on the cyanobacteria filaments because photosynthesis and nitrogen fixation cannot occur in the same cells. There are fine protoplasmic connections between all the cells in the filaments so the products of both cell types, nitrogen compounds and carbohydrates from photosynthesis, can be exchanged.

Because the manufactured products of these specialized cells can reach the other cells only through the chain of cells, it is crucial that the heterocysts be placed at regular distances from one another (Fig. 53). This is apparently achieved by the production of a diffusible inhibitor that keeps neighboring cells from turning into heterocysts; the differentiation of additional heterocysts will occur only at some distance along the chain of cells where the diffusing inhibitor becomes too weak to prevent their development. A heterocyst is created by the transformation of a normal vegetative cell; all the cells in the filament have the potential of turning into heterocysts.

Another good example of a spatially patterned organism where two cell types appear is found in the cellular slime molds. Here some cells turn into stalk cells and others into spores (Fig. 23). In this case, as I discussed earlier, there must have been an enormous selection pressure for small, stalked fruiting bodies, presumably for more effective spore dispersal; some of the cells differentiate into stalk cells to support the mass of spores. The remarkable thing about the stalk cells is that they become vacuolate and die; they are altruists and help their genetically related (and often genetically identical) sister cells to pass on their genes by forming spores. In Chapter 4 I also contrasted a slime mold species that has two cell types (*Dictyostelium*) with those that have only one in which all the cells of the fruiting body can either propagate as spores, even those in the stalk (*Guttulinopsis*), or where all the cells secrete a noncellular tube of cellulose for a stalk and then turn into spores (*Acytostelium*). In the latter case they could be considered temporal two-cell types (as in spore-forming bacteria): one would be the stalk-secreting phase of the amoebae and the other would be spore forming. (From this perspective most cellular slime molds would have three nonsexual cell types: vegetative amoebae, stalk-forming amoebae, and spores.)

In addition to these asexual structures, the differentiation of sexual cells also evolved. The question of why there has been such a strong selection pressure for sex is by no means easy to answer, as we know from the current literature, but that such pressure has persisted from the early eukaryotes onward is amply demonstrated by the fact that almost all of them are sexual. From the very origin of eukaryotic sexuality the gametes were different from one another in some of their chemical properties, and later a morphological difference developed so that the egg became large and immobile and the sperm remained small and mobile. These differences between the sex cells took on other forms in terrestrial plants, as in the fungi and in higher plants where the flagellated motility of the gametes has been lost.

The further addition of cell types in plants seemed to be a response to the need for improved supporting cells in the form of specialized thick-walled cells, and for the transport of material by special conducting cells. These major innovations are certainly ones that go hand in hand with increase in size, but there is no reason why they could not have evolved as innovations solely in response to selection pressure for increased efficiency. Even if there were no size change, a fungus or an alga which, by mutation, began to differentiate some cells as specialists for support or conduction might have acquired an advantage over less well-endowed competitors. Under some conditions of special environmental stress, the gain from these differentiations might have been crucial.

In the case of primitive multicellular animals the same argument applies: the innovations are ones that are needed for size increase, but they could well have evolved without size increase and have been selected purely on the merits of increased efficiency and the advantage such efficiency provides for success in competition. This would be the case for the flagellated collar cells of the sponge's circulatory system and for the cells forming the gut system of cnidarians. In cnidarians motility seems to have become increasingly a property of the whole organism, rather than that of individual cells. The advantage of being able to move about as a unit is highly significant. An organism such as *Hydra* can move to an area where food abounds, and once there it can grasp the food with motile tentacles. For this purpose, specialized contractile or muscle cells have arisen, and the movements of these muscle cells are coordinated by a nerve net, a group of interconnected nerve cells that have no central ganglion or brain but are capable of organizing waves of contraction (Fig. 65). This example is particularly significant because it shows a kind of cooperation between differentiated cells with which the adult is endowed for life. In more primitive forms all cooperation between cell types occurred during a transitory period of development. The very basis of all the extraordinary powers of motility and coordination which are so characteristic of the entire animal kingdom stems from this kind of permanent interaction between cell types. And as one proceeds up the phyla in the animal kingdom, in general one finds a vast increase in the efficiency of this neuromuscular system laid down in primitive cnidarians. This increase is characterized by more effective (and specialized) muscle cells, faster conduction of impulses along the processes of nerve cells, and a centralization of the nervous system. Again it can be argued that all of these complications would be expected with size increase; but as before, it is equally likely that this progression in efficiency evolved on its own merits because it improved the individual animal's ability to survive.

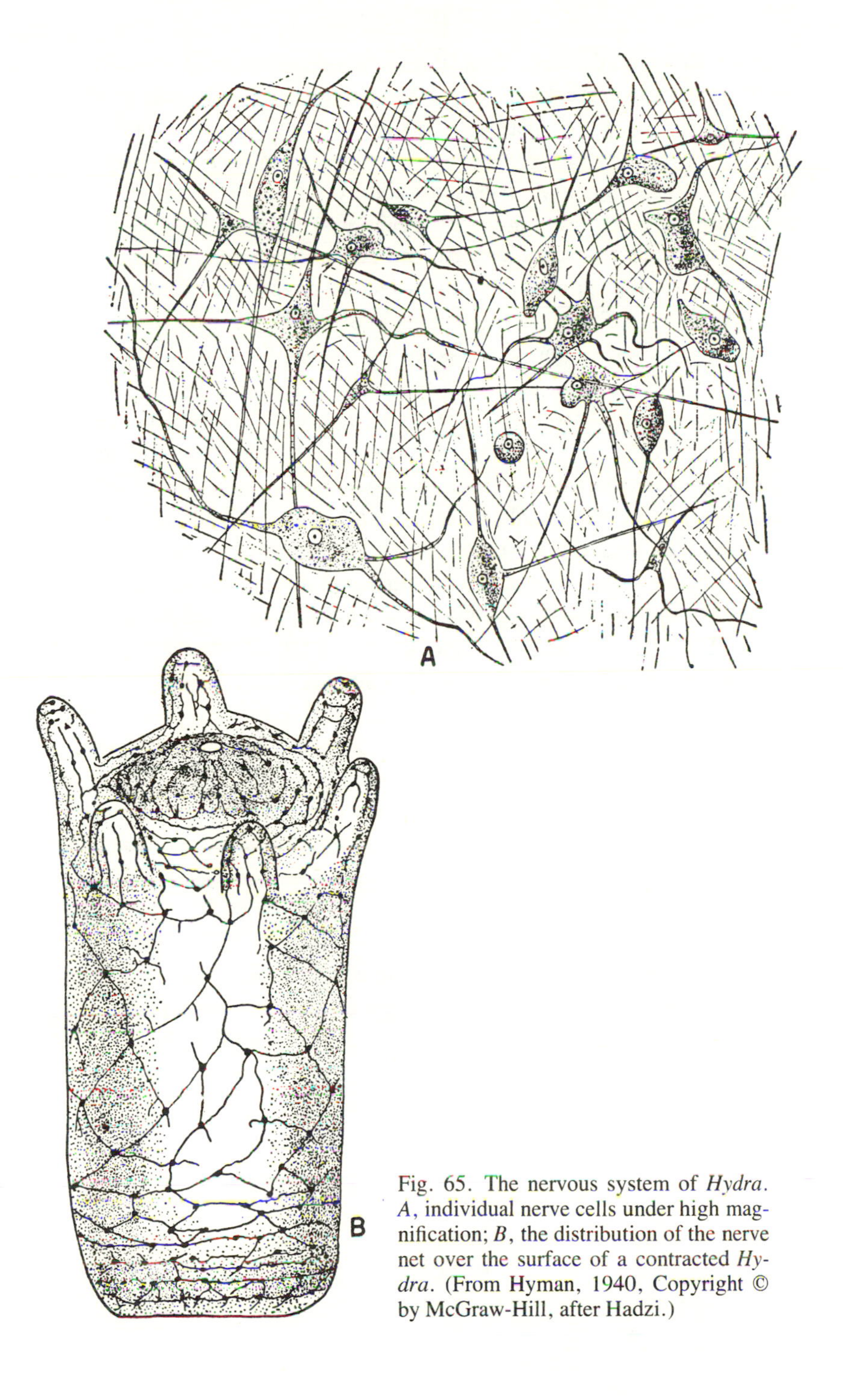

Fig. 65. The nervous system of *Hydra*. *A*, individual nerve cells under high magnification; *B*, the distribution of the nerve net over the surface of a contracted *Hydra*. (From Hyman, 1940, Copyright © by McGraw-Hill, after Hadzi.)

Yet another kind of cell differentiation is found in cnidarians that is interesting for different reasons. Nematocytes, the cells that produce small stinging capsules (nematocysts), are most concentrated in the tentacles of cnidarians and are used for defense and for paralyzing captured (but still struggling) food organisms (Fig. 66). These nematocysts are extraordinarily elaborate structures; not only do they contain poison which they inject into a foe when the cell is triggered, but they turn inside out as they spring forth. The selective advantage of these weapons is obvious; the mystery is how they evolved. Perhaps their ancestors produced only the poison that was released when the cell burst on contact. It is hard to imagine how this progressed to a complex nematocyte with its nematocyst, even with the most generous encouragement from natural selection. There are those who believe the answer might be that nematocytes came from symbiotic single-cell protozoa (myxosporidians) that do indeed resemble nematocytes. (More about this kind of solution to an evolutionary problem when we come to cellulose digestion.) But this does not resolve the final enigma in this mystery, that nerve cells and stinging cells come from the same stem-cell lineage, the i-cells. The power of this example is that it shows how elaborate and specialized cell differentiation may become, even in early, relatively simple forms.

There is another important aspect of increased internal complexity that should be emphasized: it is the complexity of the life cycle of an organism. So far I have dwelled on those organisms that have one development, but in many instances there are life histories that involve a sequence of developments. Consider, for instance, the metamorphosis from a larva to an adult, such as a caterpillar to a butterfly, or the case of some parasites that have different forms in different hosts which they invade sequentially. Each phase of these complex life cycles is a separate development, one added on to another in a time series. In other words, an increase in internal diversity means not only a greater division of labor at any moment, but a possibly changing set of such divisions over time.

Modifiability and complexity

Let us now consider the problem of how the complexity of an organism affects its ability to change. On the surface it would seem obvious that if an organism is small and has relatively few parts and few genes, it

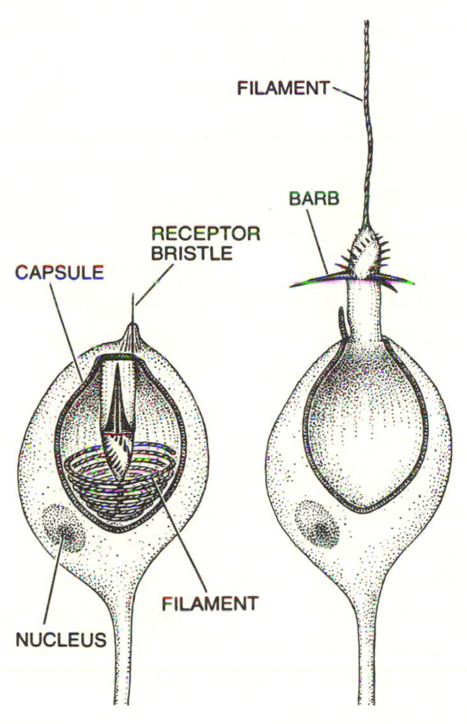

Fig. 66. Nematocytes have a highly complex structure and function. In *Hydra* there are several types. Before these stinging cells are discharged they contain a long coiled filament. When the receptor bristle is touched, an impulse is relayed to a capsule, which expels the filament (*right*). The filament releases a stinging substance, and the barb at its base helps to impale the prey, which is eventually carried to the animal's mouth. (From Gierer 1974, Copyright © by *Scientific American*.)

would be easy to affect by mutation and recombination, the kinds of changes that are needed for evolution. This conclusion is undoubtedly true, but there are some facts that give one pause. One thinks of bacteria, blue-green algae, and microorganisms in general as having existed in a relatively stable and unchanging state since their origin in early earth history, in some cases billions of years ago. On the other side of the ledger, if one looks at the evolution of mammals, there seems to have been an extraordinary variety of shapes in a relatively short time span which can be measured in millions of years (Fig. 3). The misleading aspect of this comparison is that some of those primitive microorganisms gave rise to the successful evolutionary lines that ultimately produced mammals; this is a basic fact of evolution. It means that while some microbes changed, others did not, and one presumes this is because there was simultaneous selection for both change (i.e., diversity) and stability in the environment of early microorganisms. But the fact remains that both large and small animals and plants have changed during the course of evolution, and here we want to examine the difference in how they managed this change.

One point that is often made which pertains to large organisms is that the success of a mutant gene will depend, among other things, on when it appears during development. If it appears late it has greatly increased chances of being successful; if it appears early the chances are much greater that it will be lethal and the embryo will die. This is because any genetic change appearing early will affect so many subsequent steps that it is very likely to have a detrimental influence. On the other hand, if a gene affects some superficial part, such as eye color in animals or hair density on plant leaves or stems, the differences have less chance of affecting the viability of the developing individual. Development is a sequence of steps and, in general, the later steps are dependent on the successful completion of the earlier ones. As a consequence, modifiability of the development of an organism seems to be closely affected by the degree of its complexity.

Certain characters found in living organisms appear to be extraordinarily widespread, and therefore it is assumed they are favored by vigorous selection pressure under all or many environmental circumstances. For instance, the ubiquity of sex among animals and plants argues for a general, powerful selection pressure for its retention. Even more obvious are all the pathways and mechanisms responsible for energy turnover in the organism; any failure of cell metabolism would mean instant elimination, for the selective power to retain it in working order is very strong. Less convincing cases could be made, for instance, for small organisms that tend, as we have emphasized before, to produce small, stalked fruiting bodies which bear spore masses up into the air. While this is restricted to some bacteria, slime molds, fungi, and at least one ciliate protozoan, the number of species with such structures is impressive. A somewhat less prevalent kind of adaptation is found in the form of locomotion in animals: there must have been and perhaps still is considerable selective pressure for aquatic animals to move onto land, and further for some terrestrial animals (and even a few aquatic ones) to conquer the air. And of course there have been numerous examples, from penguins to whales, of terrestrial or aerial animals that have reverted to an aquatic existence. The point is that if one could easily count the number of species that have certain kinds of adaptations one would have a very rough index of the strength of the selective forces in that direction. But this is an impractical project because, as must already be obvious in considering the examples that have been given, each case has all sorts of special circumstances that make quantitative comparisons of these adaptations only partially meaningful. There is only a rough correlation between the incidence of a trait and the degree of pressure of selection that caused it.

There is another crucial point to make. Not all of these traits arise in the same way. Some are repeatedly invented, that is, they are separate evolutionary events. This is known as convergent evolution. Others are invented perhaps only once and are retained by descent, by lineage. Flying insects, birds, and mammals come from at least three separate inventions (and possibly more among insects). The evidence is obvious for they have arisen in separate groups of organisms, in each case from unrelated nonflying ancestors. On the other hand, in the sexual system in eukaryotes, which involves fertilization and meiosis and all the highly organized chromosome activity associated with them, one can only make the reasonable assumption that this complex mechanism was invented once and was passed on to all the descendant eukaryotic organisms that exist today.

The question of how strongly a trait may be under selective pressure does not depend on whether it was invented once and passed on to the progeny, or invented many times and shows multiple convergences. But these differences have another significance which is especially relevant to our argument here. If one measured the number of convergent innovations in any one group of organisms, it is generally true that the more complex the organism, the fewer the instances of convergence. Among fungi, bacteria, and slime molds, there are a large number of independent inventions of a small fruiting body, while the invention of flying in relatively complex insects and vertebrates happened only a few times.

On the reverse side of the coin, if one thinks of a large number of the features of vertebrates, one is struck that they have a common origin and are passed on through descent. One might begin with the invention of the notochord and its development into a vertebral column; it was no doubt a single innovation that then spread into all the descendants of some primitive chordate. Other features of vertebrates have an origin among their invertebrate ancestors. In the same way higher plants, gymnosperms and angiosperms, show a great conservation of structure in the form of growth zones, structural support, conduction tissue, photosynthetic tissue, and so forth. Some of these inventions can be traced back to early ancestors among the ferns, or to the mosses and liverworts, while others, such as pollen, are characteristic only of gymnosperms and angiosperms. In the latter case they were invented in some primordial gymnosperm and passed on to all the descendants.

It is true that some inventions, such as sex and the ones connected with cell metabolism, are so ancient that they begin with the simplest organisms and transcend right through to the most complex. They appear to be such basic and essential innovations that, once achieved, they could not ever be spared. As a result we must eliminate these examples if we wish to compare simple and complex organisms, since they occur in both. However, we can rephrase the main point so that they can be included: we can say that the simpler the organism, the easier it is for it to make inventions. Some of those inventions, such as sex and cell metabolism, have been so successful that they have persisted in all their complex descendants. Others have come and gone, or become extinct in some way, but the ability to replace old ones or produce totally new inventions of broad usefulness is greatest in simple organisms, and progressively declines as they and their development become complex.

Cellulose digestion. There is an excellent illustration of the above phenomenon that is interesting not only because it reflects on the basic point, but also because it shows how the problem of making fundamental changes in complex organisms can be circumvented in evolution. Cellulose is a rich source of food, since the molecule is made up of simple sugar units (glucose) strung together in long chains. The difficulty is that the cellulose molecules are bound together in insoluble fibrils, and special enzymes are needed to break down these bundles so that ultimately the enzyme cellulase can release the glucose. Because of the complicated nature of the digestive process, plant cell-wall cellulose is unavailable as a source of energy to many animals, including ourselves. When we eat vegetables we obtain various sources of energy from them, including proteins and numerous carbohydrates such as starch. But starches, which are also made up of glucose chains, are relatively easy to digest and we can chop the large molecules up into its small sugar units. We have digestive enzymes to do so, beginning with the amylase in our saliva. But we, along with all other vertebrates, are powerless to break up the cellulose in plant fibers and it passes through us untouched as roughage.

Yet we know that many large and small mammals, from elephants to rabbits, not to mention horses and cows, survive on grass and other green parts of plants, and indeed they do manage to get the sugar from the cellulose; it is their main source of energy. They do not achieve this with their own enzymes but they have in their gut system a variety of microorganisms, either bacteria or protozoa, or both, that are capable of breaking down cellulose. The process is quite difficult and slow because of the insoluble bundles of cellulose, but these single-cell symbionts can successfully do it. The cow, for instance, has an elaborate fermentation vat in her stomach that provides a favorable environment for the microbes to process the cellulose. Furthermore, the breakdown may be interrupted by bringing the half-fermented mash back for further grinding by the back molars. When the cow "chews its cud" in this fashion, it further separates the cellulose fibers of the plant so that they can be more easily reached by the enzymes of the microbes. Other herbivores have different ways of fermenting the plant-cell walls, but the basic principle is the same and involves the help of cellulose-eating symbionts in their gut. Even termites and other wood-eating insects have intestinal fauna that allow them to devastate the wooden beams and foundations of houses.

There are many known microbial organisms that can synthesize the needed enzymes, including a cellulase. This is found among bacteria and different groups of protozoa, such as ciliates (which are found in cows) and flagellates (which are found in termites and wood roaches), and among many free-living fungi that cause wood rot. There is every reason to believe these have appeared as numerous separate inventions by these simple organisms, all under intense selection to make use of this difficult but abundant and rich source of energy. Yet if one looks at all the cellulose-eating animals, one sees only clear evidence that it has been invented in one species of shrimp and three species of the closely related isopods. All the other animals that get their energy from cellulose do so by relying on the cellulase and other enzymes of the microbes in their gut.

The moral to this story is that the invention of enzymes to break down cellulose is a relatively easy task for a simple, small organism, and the evidence is in the many cases of the independent evolution of cellulose consumption in unicellular forms, while it is extremely rare in all multicellular animals and is not found in vertebrates at all. From the point of view of adaptation, higher animals are clearly not at a loss, for many of them make use of the world's great store of cellulose by acquiring cellulase-containing microorganisms in their gut. This is a much easier way to adapt to herbivory and therefore will be favored by selection. Presumably an incipient herbivore might have taken in a certain amount of cellulose fiber, as do bears or human beings, and some cellulose-destroying protozoa or bacteria entered the gut also. They found the gut an effective place for the anaerobic (i.e., without oxygen) fermentation of cellulose and made it their permanent abode. When helping themselves they inevitably split out some extra glucose that would be of immediate value to their host when absorbed through the wall of its intestine. This sudden gift of food from the internal parasites might have given rise to a mutualism, a partnership where both sides benefit. In subsequent coevolution, the animal has in many cases altered its gut structure to facilitate fermentation, such as the multiple stomach of ruminants, which we have discussed, or the large pocket of the intestine (caecum) found in rabbits and deer. Our appendix is an atavistic remnant of that caecum; we do not ferment cellulose-eating microbes, but perhaps we had ancestors that did.

Another excellent example of such a mutualism comes from the African honey guide, a bird that can eat wax, a substance even more difficult to digest than cellulose. The honey guide will find a beehive, look for an African badger (ratel), and, by making a great fuss, lead it to the hive. The ratel eats the honey but leaves the wax for the honey guide. The bird can digest wax because in its gut it has bacteria that have evolved the appropriate enzymes to break up the long, recalcitrant wax molecules. To prove the point, these bacteria were cultured and fed to hens which for a period could subsist on wax (H. Friedman et al., 1957). The hens did so as long as the bacteria stayed, although eventually they became extinct in this foreign gut. Again in this instance the complex bird, the honey guide, has evolved an extraordinary diet by associating with the appropriate microbe.

A final example of solving by symbiosis a difficult metabolic problem in a complex organism may be found in plants. Certain angiosperms, such as legumes, do not depend solely on the dissolved nitrogen compounds in the soil, but have nitrogen-fixing bacteria lodged in their roots. The reader will remember that simple cyanobacteria solved this problem by producing two cell types: vegetative cells that were photosynthetic and heterocysts that fixed nitrogen. The reason for this separation is that the nitrogen-fixing enzyme cannot operate in the presence of oxygen, the product of photosynthesis. Higher plants are confronted with the same problem, and by housing nitrogen-fixing bacteria in anaerobic nodules in the root, they can solve the problem of producing sufficient nitrogen compounds for their growth and existence.

It would be helpful to ask why a large complex organism with an involved developmental program finds it difficult to evolve its own special sets of enzyme systems, or at least does not seem to do it frequently. Why is it so much easier to take advantage of an accidental infection of a cellulose- or wax-eating or a nitrogen-fixing microbe than to invent the new pathway itself? Producing a new enzyme in a large organism is difficult because it involves more than acquiring the right mutant genes for its synthesis. It must also be produced at the right time and the right place, for instance, in the gut cells of the animal after the gut has formed.

More generally, the genome in complex organisms is shielded by many layers of insulation, making the incorporation of novelties a formidable task. All a microorganism needs to do is produce a new enzyme and immediately it can operate on a new substrate. If it is cellulase, then the microbe can obtain glucose from cellulose. The fact that microbes also have short generations makes their mutational experiments, which produce a new enzyme, both possible and frequent. Because of their high rate of reproduction, even if many of the mutations are unsuccessful or even lethal, there are plenty of survivors. And if a new mutant with a set of enzymes that can attack a formerly unassailable substrate appears in an environment where that substrate is the main source of food, there will be exceedingly strong selection pressure for the new strain to grow and prosper. It is not surprising that the easy route of symbiosis between a microbe and a large, complex animal is the one most often followed.

One may wonder, in considering the above, how complex organisms evolve at all. They seem to have so many genes, so many multiple or pleiotropic effects of any one gene, so many possibilities for lethal mutations in early development, and all sorts of other problems due to their long and complex development. Yet obviously they are extraordinarily successful at evolving and do so with efficiency and with what might be termed geological speed. Let us now examine how this is possible: how a large complex organism can be modified to produce evolutionary change.

Gene nets. Perhaps the most important feature of the development of complex organisms is one that permits evolutionary change. It is a property of development that has not been sharply identified and defined, and here I wish to do so. I will call the phenomenon a ''gene net'' and by this I mean a gathering, a grouping of a network of gene actions and their products into discrete units during the course of development. This concept is important for it is the underpinning of the whole subject of heterochrony—the ability of organisms to change the timing of the sequence of developmental events.

Let us begin by reminding ourselves that the genes produce proteins which in time may be involved in a plethora of subsequent steps. The chain may be long, and some products along the chain will be interactive and either regulate the kinetics of the chain or communicate with some other chain or chains as inductor substances or hormones. These gene nets may also branch so that numerous substances can be produced by the release of one gene product in a particular environment of other reactants which may come from other genes. Some of these interactions may have significant developmental effects that seem quite distant from the gene responsible, that is, they are pleiotropic.

If all gene products affected all others and there was an unbroken network of interconnections, the result would be that any one gene change would affect all aspects of development. In a single-cell organism whose only development is binary fission and subsequent growth, such total interdependence of gene products might not be a great impediment. All mutations with adverse effects would be quickly eliminated by cell death, and those that were successful would have minimal secondary complications.

In a spore-forming organism with two cell types, if there is temporal differentiation the problem is solved by a long chain of steps which go from the feeding or vegetative condition to a sporulating one. At the next level where two or more cell types occur simultaneously, there must be some way of insulating the gene nets for one differentiation from those for another. This cannot be by separating them in time, for they occur at the same time. The gene nets must become isolated from one another and proceed independently. It is not total independence; they still must keep in touch with one another, for this is the way pattern formation is achieved. But the signals between them are likely to be few and quite specific. These are the signals we discussed earlier that govern the relative number of the different cell types and usually their distribution in space. But the steps leading to the differentiation of the different cell types are necessarily separate from one another and involve only rare interaction to produce a consistent pattern for the whole multicellular organism.

This general principle of the grouping of gene products and their subsequent reactions into gene nets becomes increasingly prevalent as organisms become more complex. This not only was helpful and probably necessary for the success of the process of development, but it also means that genetic change can occur in one of these gene nets without influencing the others, thereby much increasing its chance of being viable. The grouping leads to a limiting of pleiotropy and provides a way in which complex developing organisms can change in evolution.

With increasing complexity there also arises a hierarchy of gene nets. We discussed grouping associated with different cell types, but those cell types in higher plants, and especially in higher animals, are grouped into organs, which in turn may be grouped into organ systems. This is a reflection of the fact that there has been a similar hierarchy of grouping of the gene nets. On all levels (cells, organs, and organ systems), gene nets are an indispensable aspect of development of complex organisms. Only those advanced organisms with such a system of grouping will be successful in controlling and absorbing change by mutation and, therefore, competing successfully. There will be strong selection pressure for gene nets during the course of evolution. (An interesting aside: It has been pointed out to me by my former colleague Jon Seger that a properly structured "modular" computer program is put together in a way that is analogous to the gene nets postulated here.)

Heterochrony. The question is often asked, how is it possible for a higher organism to undergo a major change in the course of evolution, something that obviously has occurred repeatedly. This was dramatized for everyone with the publication of M.-C. King and A. C. Wilson's (1975) well-known paper in which they showed that even though there were relatively few differences in structural genes between *Homo sapiens* and chimpanzees, there are (at least from our egocentric view) enormous differences in the morphology of the two species. How can one produce such large morphological change with so few genetic changes?

One answer is that changes in the regulatory genes have a major effect, for one such change can greatly alter the pattern that determines what other genes are switched on or off. Another answer is that genes that affect the timing of some developmental event could mutate, and this difference in timing of some part of development relative to another could have a major effect on morphology.

This second possibility is known as heterochrony. It has a venerable history going back certainly to Ernst Haeckel, who coined the term, G. R. de Beer (1958), who brought the concept into modern evolutionary biology, and S. J. Gould (1977), who wrote a comprehensive history of the subject and has brought it up to date.

The idea is very simple: the timing of various parts of development can be dissociated so that one set of steps that preceded another in an ancestor will follow it in a descendant. To give a well-known example, some salamanders, such as the mud puppy or the Mexican axolotl, can have ripe gonads and be capable of reproduction while they are still larvae (Fig. 67). Somehow their transition from water to land, from gill to lung, has been postponed or slowed down, and their sexual maturity has either taken the normal amount of time or appeared precociously. The timing of the differentiation of the sex organs and that of the lungs seems to be entirely independent: one can move forward or backward in time without affecting the other. The prolongation of juvenile stages, as in the axolotl, is often called neoteny.

Another often-discussed example of a change in timing with impressive results is found in human beings. Compared to apes, our period of growth has become greatly extended; the whole process continues over a long time span. As a result we have an unusually long period of child-

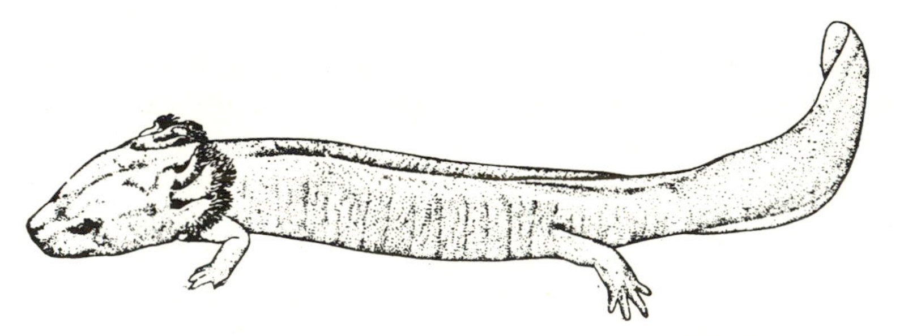

Fig. 67. A sexually mature mud puppy (*Necturus maculosus*) which still possesses the larval gills. (From Noble, *The Biology of the Amphibia*, 1954, Copyright © by Dover.)

hood (and need for parental care) and an extended period of growth of the brain; many of the structures we associate with maturity, such as a relatively flat face and partial hairlessness, are characteristic of juvenile or foetal apes, but not mature ones. This is an excellent example of neoteny and shows a way one might explain the paradox raised by the work of King and Wilson just mentioned. A few gene changes which affect timing will have enormous morphological effects, including a large increase in brain size. In fact, most of the obvious morphological differences between human beings and chimpanzees seem to be related to heterochrony. It is important to note that the growth rates of all parts of a developing child are not affected in an identical fashion, for the time at which brain growth stops, relative to body growth, is different in primates and *H. sapiens*. In the former, body weight continues to grow much longer relative to brain growth than in the latter, but otherwise the patterns are similar (Fig. 68). As a result a mature human being has a relatively large brain compared to an ape of comparable size, such as a chimpanzee.

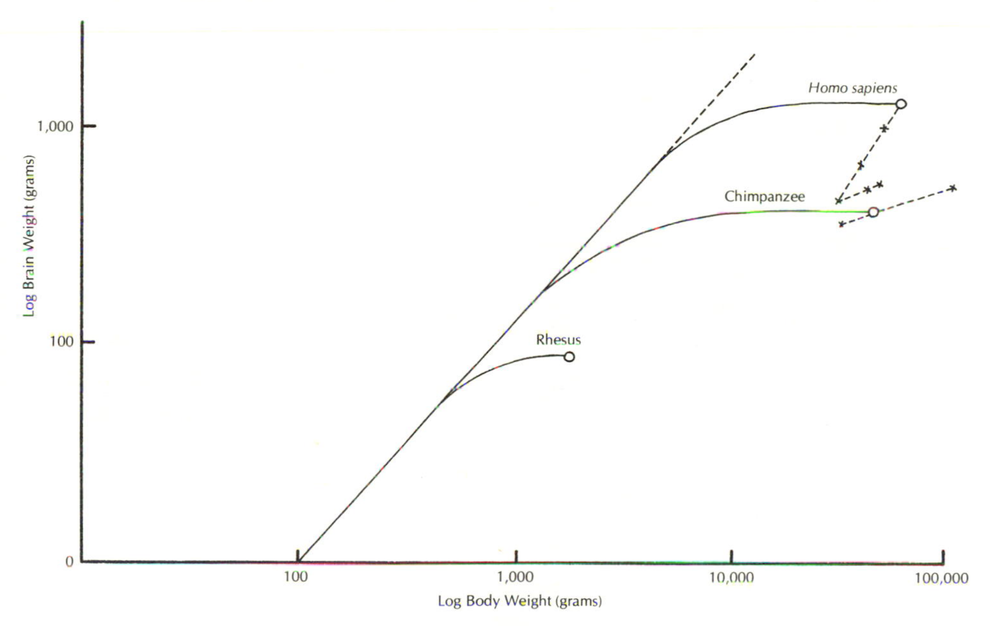

Fig. 68. The relation of brain weight to body weight during the growth and development of a man, a chimpanzee, and a rhesus monkey. (From Bonner 1980, Copyright © by Princeton University Press, after Count 1947.)

In both of these examples it is clear that some organs, such as the gonads and the lungs in the axolotl, can have their rate or timing of development shifted independently. This means that the developmental steps, beginning with the first gene products and ending with the entire organ, for the gonads and for the replacement of gills by lungs, must be quite independent from one another, and this could have occurred only if there had been a prior grouping of gene effects in the form of gene nets. This is also true in the case of brain growth and body growth in human beings and other primates. There the dissociation between the two could be achieved only by grouping the gene effects of the development of the brain. In both these examples it is fully appreciated that there may be some communication between the two parts of the organism, that is, between two gene nets, and perhaps the moment when the brain stops growing in all primates is determined by some cue given off by the growing body.

There are many other examples of heterochrony where the need for grouping the gene effects is obvious. Consider, for instance, the precocious development of the heart in the chick embryo. It forms early and serves to pump a primitive circulation for the development of the primordial mass of cells. This particular bit of heterochrony again could only occur provided all the gene messages and consequences of those messages needed for the development of the heart were grouped together in an independent unit.

In conclusion, there is an especially interesting point that can be made about gene nets and heterochrony. As complexity increases, that is, as the number of genes and the number of gene products—primary, secondary, and way beyond—increase, there is a tendency for an increase in the number of gene nets. In other words the *size* of the gene net has a limit, and if the limit is exceeded, a new net is formed. We have no idea what that size might be, nor any way of estimating if different gene nets are of similar size. One rather suspects that there is a considerable range of sizes, but it is only a guess at the moment. The important message here is that again we see a relation between size and complexity. In this case, complex development has been subdivided into units of manageable size. The advantage of such a subdivision is not only that it makes development possible, but also it permits the flexibility that one finds in evolving higher organisms; it permits heterochrony.

We have constantly alluded to natural selection as playing a role in shaping development; let us now examine a number of ways in which this has occurred. This is a different matter from how selection affects adults, although the two often overlap.

NATURAL SELECTION AND DEVELOPMENT

The idea that certain developmental stages have been modified by natural selection is a notion that was brought to the fore by W. Garstang (1922) in a well-known paper. His evidence for this came from animals in which the larval stage is often extraordinarily well adapted to its environment. It may have a special locomotory apparatus and gut system and mouthparts to cope with particular foods. The environment of the larva may be quite different from that of the adult, as in the dramatic case of insect larvae that are aquatic and the adults terrestrial (or aerial), such as mosquitoes and other diptera. G. R. de Beer (1958) has pointed out that in some instances the adults of two species of the fly may closely resemble one another but the larvae may differ radically (Fig. 69). This has been called "clandestine evolution" by de Beer, for indeed if one merely studies adults, as some taxonomists are wont to do, these great larval differences will not be noticed.

Plants have similar adaptations, and a comparable invention is that of the seed leaves or cotyledons of angiosperms. They are rich sources of nutrition and provide food until true leaves are able to obtain energy by photosynthesis. Developmental adaptations may also be seen in the leaves that follow in the young seedlings of some plants which may differ quite strikingly from the mature leaves. For instance, in some species of trees the first leaves are enormous and serve as effective light catchers in the understory, but as the young shoot reaches upward into the light the leaves become smaller (Fig. 70). Here, as in insect larvae, each of these special juvenile features helps to make the organism viable and well fed, leading to the ultimate reproductive success of the adult.

In animals the equivalent of cotyledons is the yolk. Some eggs, such as those of a bird, have an enormous amount of yolk, while others, such as those of a mammal, have a very small amount. The reason for this difference is fairly obvious: bird development requires a long incubation period in the shell without exogenous food, while a placental mammal needs only enough yolk to last until the stage where it is implanted on the wall of the uterus. One assumes that the eggs of birds and the placentas of mammals have arisen and have been maintained by natural

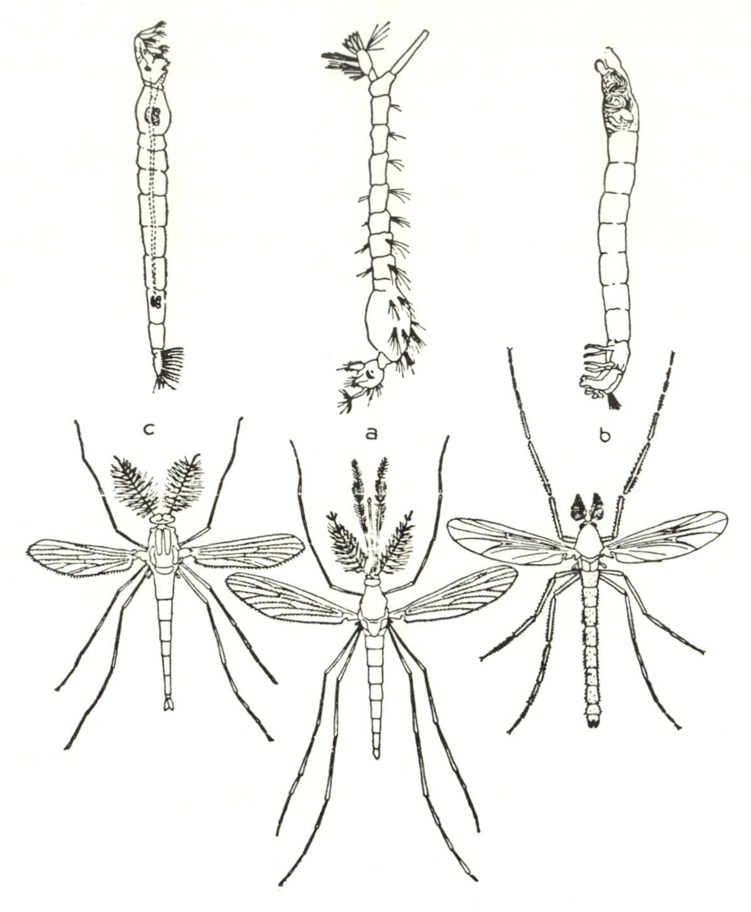

Fig. 69. An example of clandestine evolution. Note that the adults of these three flies are quite similar, yet the larvae differ considerably. a, *Culex*; b, *Chironomus*; c, *Corethra*. (From de Beer 1958, Copyright © by Oxford University Press.)

selection, and it seems quite reasonable to assume that the amount of yolk is adjusted to need in both cases.

Consider now another property of the embryo that varies in different animals: mosaic and regulative development. Superficially it seems strange that different groups of animals should have such totally different kinds of development. In one, all the cells are the same, for they are interchangeable (regulative), while in the other the fate of a cell has become fixed and remains so for long periods of time (mosaic). The

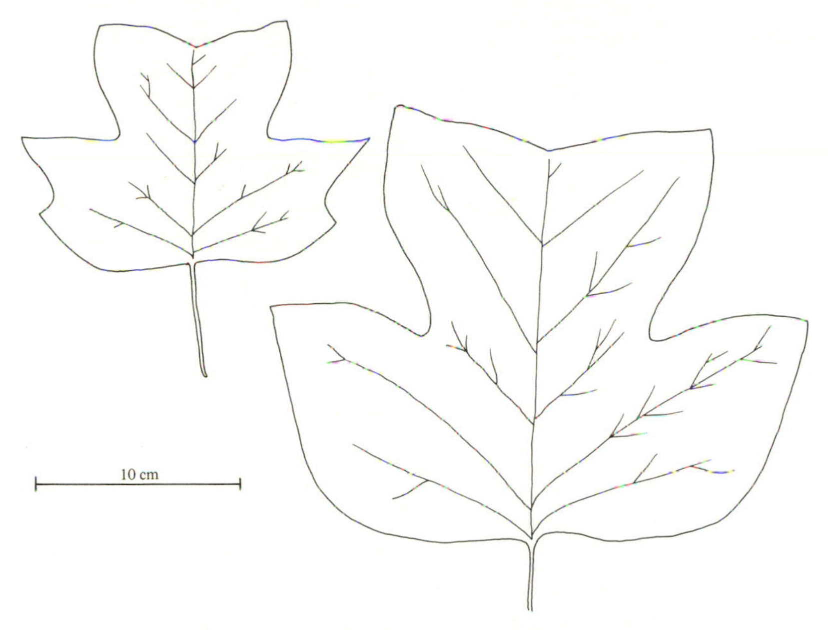

Fig. 70. Size difference in a leaf from a small sapling (*right*) and a large mature tulip tree (*left*).

main difference between the two is that although the cell cytoplasms appear to be approximately equivalent in regulative development, in mosaic development there are key macromolecules which are segregated into specific cells. Because the molecules are large they cannot easily move from cell to cell and therefore only influence cell differentiation in a restricted region.

The properties of these two systems are very different. Mosaic development, where there is a strict cell lineage of traits, has the advantage of simplicity and perhaps the reduction of developmental errors. If the macromolecules responsible for differentiation are put in the right cell compartments initially, all that is needed from then on is to control the cell divisions which in turn control the proportions, the localization, and the pattern that will follow. On the other hand, mosaic development is ill suited to recover from any kind of injury or disturbance. Here lies the advantage of regulative development. By keeping its cell differentiation options open until late in development, the embryo can be bisected or damaged in various ways and the fragments will recover. Consider, for instance, the case of *Hydra* or related cnidarians. The cells of the first cleavages of the fertilized egg can be separated, even up to the sixteen-cell stage, and each one will produce a normal, diminutive larva (Fig. 71). The larva can be cut and each fragment will reorganize. One can even fuse larvae to produce a unitary giant larva. Once *Hydra* or any other hydroid assumes its mature form with a mouth and tentacles, it still can be cut up into fragments and each one will grow into a new adult. It is even possible to reduce an individual to a rubble of cells by mincing it with a fine knife, and the cell mass will reorganize into a new hydroid.

Perhaps these aquatic organisms, which are exposed to predators and the action of violent water currents, find such remarkable powers of reconstitution a large asset in self-preservation. It is possible that they could never survive and reproduce were they not so highly regulative. Unfortunately these are speculations; there is no obvious way of testing such hypotheses. The case of *Hydra* is particularly instructive. Remember that it has certain cells that are mosaic, namely the stinging cells (nematocytes) and the neurons, both of which come from the stem cells, the i-cells. Neither a neuron nor a nematocyte can change into anything else; they are terminally differentiated. All the other cells which line the gut (endoderm) and the outside covering of the animal (ectoderm), such as gland cells, muscle cells, and epithelial cells, can apparently change from one type to another during regeneration. For instance, it is possible to regenerate a perfect hydroid from isolated ectoderm or isolated endoderm cells. Therefore, in *Hydra* there are two classes of cells: those that are terminally differentiated and are the mosaic progeny of i-cells, and all the other cells which remain embryonic and retain the power of regulation. With this dual or hybrid system, *Hydra* can simultaneously produce differentiated cells that are so specialized they are no longer capable of further cell division or development, while all the other cell types maintain that capability permanently. If any of our evolutionary speculations have any substance, one can argue that the mosaic system of cell lineage is the most efficient way of replacing the cell types that become terminally differentiated, and for all the remaining cell types it is most advantageous to retain the flexibility of regulative development.

Many of the truly mosaic forms, such as ascidians, are also aquatic. Perhaps they have other ways in which they can protect themselves. It was pointed out earlier that some mature ascidians regain their powers of regeneration as adults. These species might possibly be relatively protected during their embryonic and larval stages, but as adults they

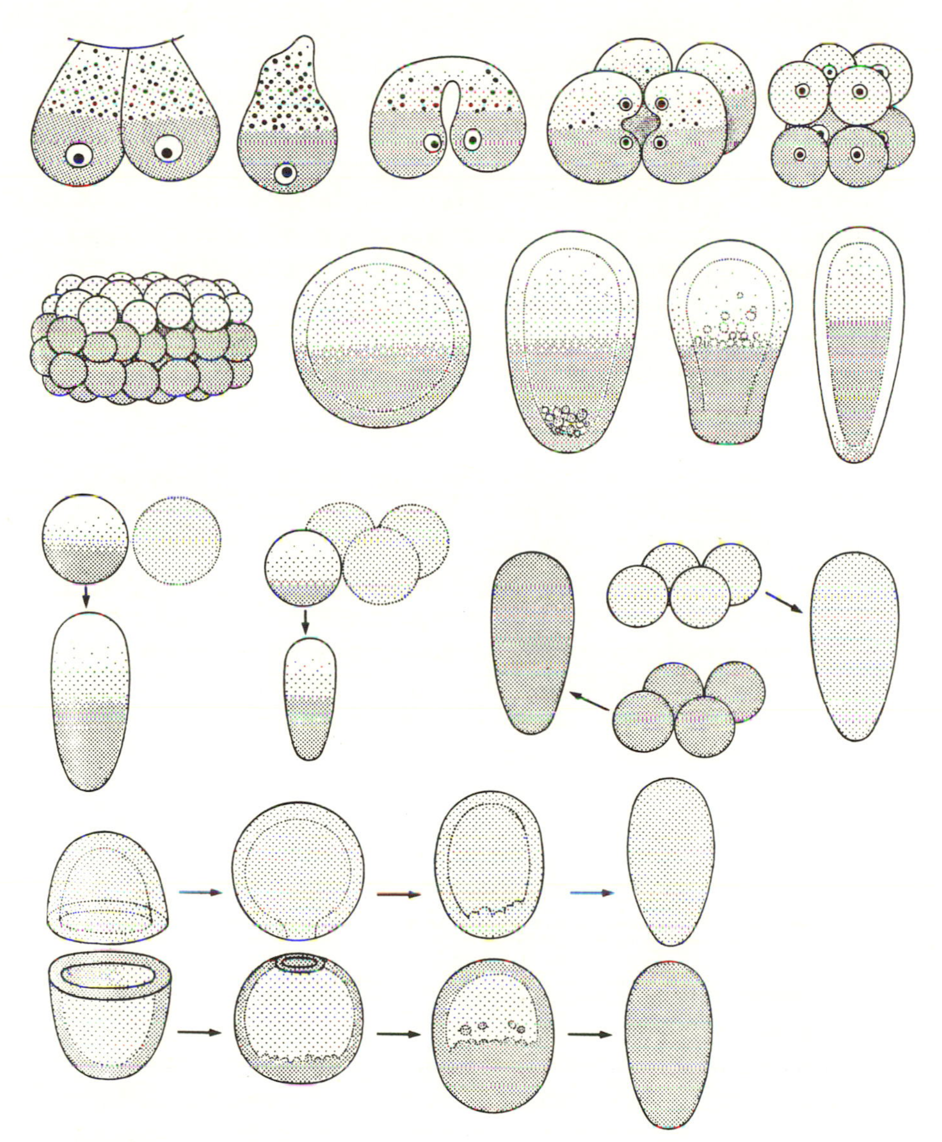

Fig. 71. Normal development and development after isolation of parts of the embryo in the hydroid *Sertularia*. *Top two rows*: Normal development showing cleavage and the formation of a planula larva. Note that the pigment at the animal pole in early cleavage becomes the internal endoderm in the larva. *Bottom two rows*: Here, whether one of two blastomeres, one of four, of eight, or a half of the whole blastula is cut away, each isolated portion gives rise to a normal, diminutive larva. This is so even when the pigmented cells are separated from the unpigmented ones, indicating a totally regulative development. (From Mergner 1971, Copyright © by Elsevier, after the experiments of Teissier, 1931.)

are more susceptible to predation or wave action, or both. Also, many adults are clonal and reproduce by budding, a process that would require regulative powers in the adult. There are a number of reasons why regulation in adults might be selectively advantageous.

In conclusion, if we look at the variety of different kinds of development, such as mosaic and regulative development, we can postulate that each, depending on the environment, has certain advantages. But this remains highly speculative, and it is also possible that the difference between these two modes represents alternative but equally successful mechanical ways of making embryos more complex as they increase in size.

Cell competition. An important new set of ideas bearing on the relation of natural selection to development comes from the recent work of L. W. Buss (1982, 1983, 1987). His basic argument is that there could be, beyond selection at the level of individual organisms, a selection as a result of competition between genetically different cells within a developing organism. Any mutational change in a developmental cell lineage will, however, die with the organism and not be passed on to the progeny, for it will not be represented in the genes of the egg or sperm. This is August Weismann's notion of the separation of the generative or reproductive cells (germ line) from the cells of the body (soma) and it only applies generally to higher animals. Because Weismann's idea fits so well for complex animals, it has become a basic assumption for modern population genetics, which is concerned with the changes of the gene constitution of the germ lines of individuals in a population. As Buss correctly points out, for many years the critics of Weismannism have shown that in higher plants, many invertebrates, and most lower forms such as algae and fungi, it is commonly possible to obtain a whole new individual from one somatic cell. There is no separation of the germ from the soma, and if this somatic cell has a mutant gene, it may well be passed on to the progeny. The consequence is that all or at least many of the cells can compete to become the ones capable of propagating the next generation. The most successful form of competition would be for a mutant cell type to become a sexual reproductive cell; in this way it will be sure to have its new genetic constitution passed on to the next generation.

There are a number of different forms this somatic-cell gene variation can take. The mutant can become extremely deleterious, as is the case in mammalian cancer. Because cancerous cells have acquired the trait of rapid growth, and further are capable of metastasis (that is, spreading to other parts), they are competitors that gain a Pyrrhic victory; for while they take over the body, the body dies and they may not be passed on to the next generation. It should be added that in mammals there is a germ-line genetic connection with these somatic mutations. It is well known that all sorts of tumors in human beings occur at a much greater frequency after forty years of age. There must be a strong selective pressure against the appearance of early cancers, for stricken individuals are less likely to reproduce and pass on the deleterious genes; those with late cancers perpetuate such genes because their illness hits them largely after they have had their children. In some way germ-line genes can control the timing of these deleterious somatic mutations, a further illustration of how cells within a complex, multicellular organism can communicate with one another in remarkable ways.

As Buss makes clear, higher plants do not have the same difficulties with tumors, for their cells are encased in cellulose walls that prevent cell movement and metastasis. Therefore, their tumors are not lethal and appear as great bulbous growths on a trunk or a branch. He goes on to show that favorable mutations in somatic cells of plants have an excellent chance of succeeding, because so many plants propagate by vegetative means, for instance by budding from spreading roots. Furthermore, a whole shoot of a plant may be the result of a mutant cell, and if that shoot flowers it will be able to carry the new variant in its sexually reproductive cells. Unlike mammals and other higher vertebrates, higher plants (along with many invertebrates) do not sequester their germ-line cells at an early stage in the truly Weismannian manner; contrary to his doctrine, somatic mutant cells may give rise to changes that will be passed on in the generations that follow.

Beyond such instances where the somatic mutation becomes deleterious and destructive, or beneficial and passed on, Buss discusses all sorts of intriguing possibilities where a somatic change will make for increased complexity within the individual. First, I will discuss his own example of a case in which this has undoubtedly occurred, and then I will go on to his more general argument of how cell-type competition within a multicellular organism could lead to the sequestering of the germ cells and to cell differentiation in general.

The example arises from the work of M. F. Filosa (1962), who showed that a laboratory stock of a cellular slime mold (*Dictyostelium mucoroides*) contained two cell lines which could easily be isolated and grown alone. One gave the normal wild type appearance, while the other was highly abnormal and appeared to be defective in its ability to form proper stalk cells. If these two cell types were mixed together again in any ratio from 90 percent mutant to 10 percent normal or vice versa, the fruiting bodies appeared normal. Furthermore, after a generation, regardless of the ratio of the initial mixture, the ratio in the fruiting bodies returned to a constant 10–15 percent mutant and the remaining cells normal. In order to fully appreciate what is happening here we must remember that these are aggregative organisms, so that it is always possible for cells of different genetic constitution to coaggregate and become part of a single, multicellular fruiting body.

In thinking about Filosa's experiments, Buss (1982) asked whether or not this might be a general phenomenon in slime molds in the wild and not just a peculiarity of one laboratory stock culture. To test this he made a bundle of micropipettes, plunged them into some soil, and then carefully separated the contents of the pipettes so that he could record which cells were found close to others. From these pipettes he produced many cultures from single cells and found instances of mutant cell types intermixed with wild type, and, as with Filosa, these mutant types failed to make a proper stalk. From this finding Buss made the most interesting suggestion that these mutant cells are parasites of the normal cells. They cannot make stalk, but by combining with normal cells they can succeed in being propagated from the communal spore mass. In order to achieve a stable relationship with its host, a parasite must devise a method to avoid taking over completely; the mutants which Buss found in the soil also had the property of remaining at a low percentage of abundance and maintaining the ratio over a series of generations.

This phenomenon raises the intriguing possibility that in any organism autoparasites can appear. They have two requirements in order to be successful. One is that they show restraint and do not eliminate their host, and the other is that they manage to pass on into the next generation. In cellular slime molds they do this by entering aggregates of normal cells during asexual reproduction. In organisms that stem from a fertilized egg they must manage to enter their genes in the germ line. This is easy for those organisms in which the germ cells may arise from any cell in the body, as happens in many invertebrates and in all plants.

One of Buss's interesting arguments is that once there is a population of different cells within a multicellular organism there will be a competition not only for entering the germ line, but also, once that is accomplished, to sequester the germ line cells (thereby excluding the somatic cells) so that their ability to contribute to subsequent generations is made even more secure. One wonders how this might be accomplished, and let me give two examples in lower forms where one sees steps that represent the beginning of a restriction of a germ line. The first example is the case of two genera of slime molds, members of a group (acrasids) that have apparently evolved independently and are convergent to the slime molds we have discussed above (dictyostelids). One acrasid, *Guttulinopsis*, after aggregation produces a fairly crude stalk with a mass of spores on top (Fig. 72). The other, *Acrasis*, develops in much the same way although the fruiting body branches into chains of spores. In *Guttulinopsis* all the cells, including those in the stalk, are capable of starting new colonies; but in *Acrasis* the stalk cells have lost the ability to propagate. We might assume that in *Guttulinopsis* there are two competing kinds of cells: those that tend to appear in the stalk, and those in the spore mass. Our next assumption is that during the course of the evolution of a descendant such as *Acrasis* (also an assumption), the

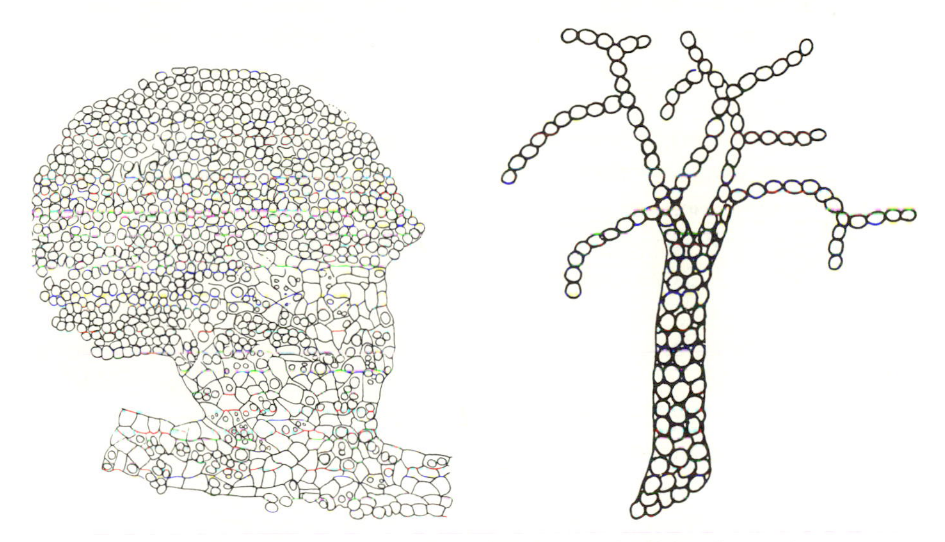

Fig. 72. Two different species of acrasid (a group of cellular slime molds). In *Guttulinopsis (left)* the stalk cells are capable of propagation and in this respect resemble the apical spores. (This is a section through the fruiting body.) In *Acrasis (right)* only the terminal chains of spores are viable; the stalk cells are not viable. (From Olive 1965.)

spore cells have somehow managed to suppress the generative ability of the stalk cells. The *Acrasis* spore cells possess the genes for the stalk cells too, but there is some mechanism, perhaps initially an inhibitor from the spore mass, that seals the fate of the stalk cells. In this way one has progressed from one to two cell types.

The other example is from *Volvox*, the colonial green alga. It consists of a globe of some two thousand cells, and in asexual development only a few of those cells (gonidia), which are found in the southern hemisphere in the colony, are capable of developing into daughter colonies (Fig. 54). However, one finds in smaller, more primitive colonies of the volvox line that any of the cells of the colony can produce a new colony. It would seem that *Volvox* has evolved from one to two cell types and in this way separated the germ line from the soma. Here we do not have an aggregation of cells as we did in the previous example; one must assume that, if we are to give this evolution a Bussian interpretation, a mutation appeared which resulted in the products of a certain number of cell divisions turning into another cell type, a gonidium. These gonidia remain large cells while the other cells continue to divide and produce the vegetative cells. However, those vegetative cells become flagellated and are usually incapable of further division; in this they resemble the stalk cells of *Acrasis*. Fortunately, the story does not end here, for R. C. Starr made the most interesting discovery of a mutant of *Volvox* in which, like its simpler and smaller relatives, all the cells of the globe are capable of starting a next generation. One mutation wiped out the supremacy of the gonidia, and they no longer are the only cells that can reproduce new colonies asexually. One might imagine that normally the gonidia put out an inhibitor preventing the vegetative cells from further division, and that the mutation is either a reduction of the inhibitor or a reduction of the vegetative cells' sensitivity to it. Whatever the mechanism, here we have two cell types appearing where one was present before, and apparently the genetic difference between the one and the two cell type condition is small. Once there were two cell types, however the second one might have arisen, it alone retained the power to pass on genetic information to the next generation. Clearly, information for both the vegetative cells and the gonidia must be in the genome of all the cells. This could have arisen as a somatic variation in the incipient germ line cells and would then automatically be present in all cell types. The novel genes could either be governing an inhibitor system (a regulative explanation), or there could have been a cell-cycle mutation that caused one cell line to remain as gonidia and the other to divide further to become vegetative cells (a mosaic explanation). This is simply to say that the competition between cells proposed by Buss, which produces a kind of order by cell dominance, could be achieved either by a mosaic or cell lineage means, or by a regulative means involving inductor substances passing from one group of cells to another. In either case *Volvox* is a helpful example, for it provides an insight into the possible origin of a division of labor among cells.

Besides cells competing within a multicellular organism, there has been considerable interest of late in the genes themselves competing and becoming the ultimate units of selection. The case was stated with great clarity by R. Dawkins (1976) who showed the merit of thinking of each gene as a competitor and that, in its struggle for what he called its selfish interests, the whole organism benefits. The organism is what he calls the vehicle, for it is the way the replicating genes ensure their perpetuation. So the genes specify an individual organism (a phenotype) that is efficiently designed to ensure successful reproduction. In a later book Dawkins (1982) extended the idea to show that the effect of genes not only reaches their own phenotype in its development, but also indirectly affects other phenotypes that interact, as is evident in all forms of symbiosis, competition, and even predation. Therefore there is a great spectrum of levels in which one can consider the effects of natural selection: the level of the gene, the cell, the multicellular individual, the level of interaction between individuals of the same species, and finally between individuals of different species. Furthermore, it should be emphasized that the trends towards increased size and increased complexity are, as we have seen, evident at all of these levels, and no doubt this is the main reason why they are primary evolutionary trends.

Social insects. Another level of complexity is that of social organisms. It is particularly instructive to examine the development of a colony of social insects, for many of the mechanisms which are evident for the role of cells within a developing organism find their parallel in the interaction between individual insects within a developing colony.

In social insects the information necessary for the capabilities of the different castes of neuter workers and reproductives are all contained in the genes of the fertilized eggs of the queen. This includes the ability of any individual larva or nymph to respond to different environmental cues, especially nutritional and pheromonal ones, and, as a result, to become a particular size and shape. We must ask what kinds of competition exist between the different castes of a colony and to what extent they can be interpreted in the terms which Buss applies to cells in the development of individual organisms.

In the first place, it is clear that the queen outcompetes all the workers to become the individual who carries on reproduction; she is the social equivalent of the germ line (W. M. Wheeler 1911). This is true of the males too, but they participate little in the economy of the colony. The queen ant is the one who, by regulating the amount of food received by the larvae or by producing inhibitor pheromones, ensures that she is the only reproductive member in the colony among all the neuter (but genetically female) workers and soldiers. In termites the workers are genetically both male and female, and they are also inhibited by pheromones from being sexually active. In both cases the ''germ line'' is sequestered in this regulative way. Note that the idea of inhibitors, which was put forth as a hypothesis for *Acrasis* and *Volvox*, is a reality in social insects. Also, in a sense the queen becomes a parasite on her own neuter offspring, for they do all sorts of vital chores from which she is relieved so she can concentrate on propagating the germ cells. There is a fine line between parasitism among members of the same species and a division of labor.

When we look at the gamut of increasing complexity that has evolved over geological time and how it has been molded by natural selection, and how it can be put together so that it works mechanically from generation to generation, we cannot avoid a feeling of awe. But we should not let this awe overcome our reason. It is crucial to examine these problems as phenomena that might have microexplanations. I cannot assert that they do, for that is asserting the unknown, but it is a sensible position to adopt. This kind of reductionism, beginning with Darwin and Mendel and leading to modern molecular biology today, has been our most successful approach for well over a century. But in looking for those microexplanations we should never lose sight of what we want to explain. That is the main purpose of this chapter, and indeed of this whole book.

Further References

New thoughts on the distinction between clonal and aclonal organisms are comprehensively reviewed in J.B.C. Jackson et al. (1985). Estimates of the number of cell types in different groups of organisms may be found in Bonner (1965), S. A. Kauffman (1969), and R. A. Raff and T. C. Kaufman (1983). The length and weight of record large kelp are found in T. C. Frye et al. (1915). Some of the important early work on the amounts of DNA in different organisms was by A. H. Sparrow et al. (1972), and the subject is well reviewed in R. A. Raff and T. C. Kaufman (1983). My discussion of the three constructive processes of development may be found in Bonner (1952, 1974). A recent statement of G. S. Stent's views on the relation between genetics and development may be found in Stent (1985). R. J. Greenspan (1988) has a most interesting paper that bears on gene nets and complexity. Basing his work on studies of specific mutants, he defines four classes of genes which to varying degrees are regulatory and operate on different levels of complexity within the developing organism. The more recent generation of patterning models following A. M. Turing (1952) are those developed for reaction-diffusion systems by A. Gierer and H. Meinhardt (see Meinhardt 1982) and those developed for mechanical models by G. F. Oster, J. D. Murray, and G. M. Odell (see Oster et al. 1983). The work of C. N. David and R. D. Campbell on cell differentiation in *Hydra* is reviewed by A. Gierer (1974). For a fuller discussion of the origin of the egg cytoplasm in the evolution of life cycles, see Bonner (1965, 1974). A summary of the evidence for neuronal cell death in the development of the nervous system may be found in W. M. Cowen et al. (1984). The work of M. Lüscher on the effects of pheromone on morphology in social insects is reviewed in E. O. Wilson (1971: 188–194). For a discussion of all aspects of heterocyst formation in cyanobacteria, see C. P. Wolk (1982). A useful discussion of the complexities of cellulose digestion may be found in D. E. Aiken and F. E. Barton II (1983). The early work on the development of different members of the *Volvox* line is reviewed by A. Kuhn (1971: 112–118). The more recent work on differentiation mutants of R. C. Starr and others is reviewed in G. Kochert (1975).

Chapter 7

Animal Behavior: The Pinnacle of Biological Complexity

The nervous system ▪ Behavior ▪ Natural selection and behavior ▪ Conclusion

THE NERVOUS SYSTEM

One of the greatest accomplishments in the evolution of development is behavior. I have already pointed out that in order for genes to produce a large nervous system, a special mode of development was invented during evolution. Instead of all the neurons and their connections in the genome being specified, certain general parameters are specified which involve the number of neurons, their tendency to form synaptic connections, and their distribution or pattern in space. With the beginning of function, of behavior, those neurons and the connections between neurons that participate in the functioning remain stable and become part of the nervous system, while those that do not, wither away. In this fashion it is possible, through a gene-directed process, to produce a structure far more complex than anything the genes might have specified in detail. In this chapter we will see that there is yet another dimension to this complexity, for behavior itself not only arises from a nervous system based on a higher order of complexity because of its mode of development, but behavior as a process can attain an even higher level.

The structure and the functioning of the nervous system is a vast subject that could easily occupy a great deal of our time, but here it is my purpose to choose those few outstanding features that illustrate my argument. To biologists the initial points will seem very elementary, but I urge the reader to bear with me as I gather the main components together to show in what way the complexity of behavior has arisen and then go on to discuss its properties and peculiarities.

The primordial characteristic of nervous systems is that they respond to their environment. For this they have special cells called receptors.

These receptors have specific macromolecules (receptor molecules) in them that react to particular environmental cues. The kind we have discussed before, and an especially important one in the nervous system, is a receptor cell which responds to specific chemicals in its environment. A good example would be the taste buds in our tongue. Some of these receptor cells respond specifically to sweet tastes (others to salt, sour, or bitter tastes). There are proteins on the cell membrane that can attach to a general class of substances, such as sugars of many kinds; and when the receptor protein-sugar chemical combination has taken place on a sufficient number of receptor molecules, the whole cell becomes depolarized and can send a signal to an attached nerve cell. What is meant by depolarization is that the cell, which has been carefully pumping certain ions in and others out (ions such as sodium and potassium), suddenly opens its pores and the ions rush through and come into equilibrium so that the concentrations inside and outside are the same. Then the cell slowly repolarizes and pumps the ions either in or out, depending on the ion, and the cell is ready to receive new sugar molecules should they pass again into the mouth. Depolarization will pass as a wave down a nerve; this is the nerve impulse. Therefore, in the receptor cell a very specialized device receives an external signal and converts it into a simple message that runs along a nerve.

The principle is that for each kind of stimulus there is a special kind of cell that responds only to that stimulus. Among chemical receptors there is a division of roles, since not only does the tongue have the four basic tastes (sweet, salt, sour, and bitter), but in our nose we have many other receptors which permit us to discriminate with great subtlety many different nuances of aroma. The blending of these signals in the brain helps to identify a particular perfume or appreciate the fine French sauce on the fish.

The other kinds of stimuli for which animals have receptors are all the physical forces that surround us: light, temperature, pressure, sound, and even electricity. Not only can animals sense light from dark, but many have eyes which can see whole images, and some species see them in beautiful color. In the retina of our eyes are special pigments which are sensitive to certain wavelengths or intensities of light. There are four of these, three types for color and one for low-intensity light. We respond to temperature with two kinds of receptors in the skin: one responds to heat and the other to cold. For pressure we have special receptors in our muscles and another kind under our skin, so that if we press against an object in the dark we can feel it. In our ears we receive sound and also transform it into nerve impulses. Sound waves are pressure waves, so hearing is closely related to feeling. The sound penetrates into the inner ear, the snail-like cochlea, and, depending on the wavelength of the sound, will cause the vibration of hairs in a specific region of the cochlea. The range of our sensitivity has strict limits and may be different from those of other animals; for instance, we cannot hear the high-frequency noises that a bat uses to find and catch prey in the dark. Some animals have a very limited set of sound-receiving capabilities; crickets or frogs can hear only the calls of their own species and are deaf to most of the other sounds that surround them. Relatively few animals have receptors to record electrical signals. Human beings do not, but many fish use them to orient in murky water, or find prey or even mates. Not only is the electric eel capable of giving off electric current, but many other fish do so constantly in less spectacular (in fact minute) amounts. These can be sensed by other fish that have special receptors in the skin, and in this way they have yet another efficient means of communication.

The nervous system is so arranged that the organism can respond to a stimulus received. Unless it is a simple reflex, the nerve impulse first goes to the central nervous system and then out to an effector. The most common type of effector is a muscle. If, for instance, we touch something that is burning hot, our hand retracts quickly. The heat receptors, and maybe the special pain receptors, have sent a message and the response is a contraction of various muscles to pull the hand away. Besides muscle movement there is a variety of less common effectors. For instance, a firefly will respond to the flash of a potential mate by emitting a flash of its own. The nerve impulse is transmitted to certain cells that mix key chemicals to give off the bioluminescence. This comes under the general category of secretion, and secretory cells are another kind of effector.

The most interesting aspect of the nervous system is undoubtedly the central processing between the input from the receptor and the message sent on to the effector. There are two aspects we shall consider here: the evolution of the central nervous system, and the fashion in which one neuron passes information on to the next. Let us begin with the second.

The nerves in the central nervous system are almost entirely connectors between nerves or interneurons; it is only at the periphery of the system that one finds the connections to receptors or effectors. The great majority of these interneurons connect with one another by synapses. These are remarkable one-way valves that work in the following way. When a nerve is stimulated to form an impulse which passes as a wave of depolarization down its fiber, it will eventually reach a bulbous end that lies close to another nerve but is separated from it by a narrow gap (Fig. 73). The depolarization, when it reaches the bulb, normally cannot by itself cross the gap, but it causes the release of a small signal molecule in the gap by opening up minute vesicles full of this "neurotransmitter" substance which diffuses across the gap and on to the opposite side. In the membrane of the other neuron are special receptor molecules that can grab that particular signal molecule. When a sufficient number of these receptors are filled, the second neuron becomes depolarized and starts a new wave, a new impulse, down its fiber. Note that in this instance the neurotransmitter is produced by only one of the neurons on one side of the synaptic gap, and the receptors are found only on the other side (Fig. 73). This anatomical property ensures that the signal can only go one way—that it is indeed a one-way valve.

Two of the commonest transmitter substances are acetylcholine and adrenaline (or epinephrine). Part of our nervous system uses one, and another part uses the other. They are both small molecules and this presumably facilitates their movement across the gap, for small molecules diffuse much more rapidly than large ones. The receptors are quite specific for the particular neurotransmitter and will only combine with it alone. There is also another interesting phenomenon: the excess neurotransmitter may be removed by a special enzyme so that the slate is

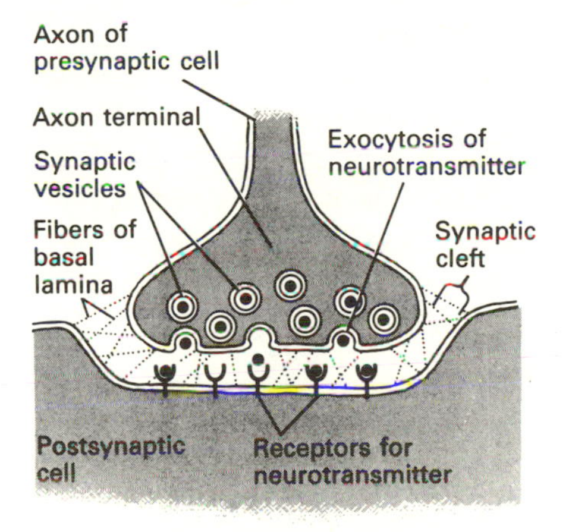

Fig. 73. A schematic diagram of a typical synapse. An electrical signal arriving at the axon of the presynaptic cell triggers the release of a chemical messenger (neurotransmitter substance), which crosses the synaptic gap and is captured by the receptor proteins on the postsynaptic cell, which causes it to initiate its own impulse passing down its axon. (From Darnell, Lodish, and Baltimore 1986, Copyright © by W. H. Freeman.)

cleaned and the synapse can operate again for the next impulse. For acetylcholine there is the enzyme choline esterase which serves precisely this function by destroying the excess acetylcholine.

One of the exciting recent areas of research on the central nervous system has been on the nature of other neurotransmitters, and it has turned out that there are many of them. One important group are small peptides, which are a series of amino acids strung together as though they were a short portion of a protein molecule. Since peptides are coded for directly in a gene, or in a fragment of a gene, it is a relatively easy task to produce a variety of these neurotransmitters. Presumably there is an equal number of specific receptors to match them, in this way greatly increasing the possibilities of specific and selected pathways in which impulses can travel in the brain. There is also another kind of ''brain hormone'' which seems to be abundant. Instead of just bridging a synaptic gap, these neurohormones are released in a general area of the central nervous system. They do not transmit directly but somehow affect the behavior of synapses that do. It is a way of modulating the response.

But the possible permutations are much greater. Any one cell can receive many synaptic connections (Fig. 74). Some cells in the brain are known to have as many as ten thousand. Furthermore, these synapses may have different transmitters with different messages: some may be inhibitory and some stimulatory. By changing the ratio of stimulatory signals to inhibitory ones, one can produce an enormous variety in the strength of the final message. Also remember that if there are so many interconnections, so many different neurotransmitter substances, giving different signals, sometimes one canceling the other, the possibilities for sophisticated processing of the information which comes in from the receptors, and the possibilities of producing the appropriate shades of response in the motor system, are enormous. In terms of complexity, this functional system is quite extraordinary and is without doubt one of nature's greatest inventions. But it did not appear full-blown as I have described; it had a long evolution of increasing complexity which I shall now briefly outline before we get on to the subject of what the system can do besides remove hands from a hot stove; that is, we will discuss the nature of behavior.

In the evolution of the nervous system there is a minor and a major theme. The minor one is that the nerves are well adapted to pass information more rapidly than can be done by merely signaling from one cell

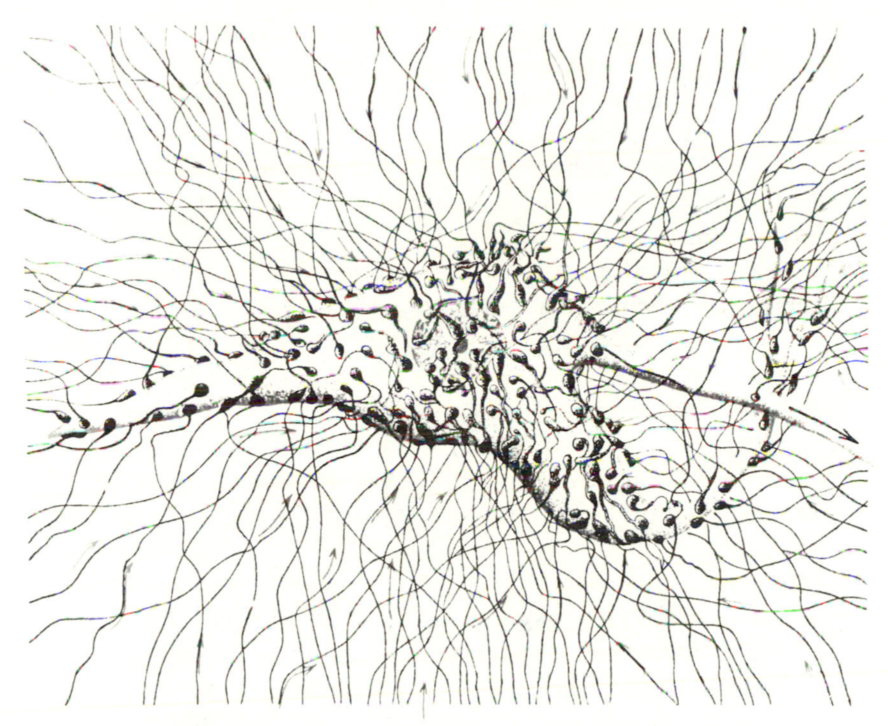

Fig. 74. A motor neuron to which many other neurons are connected by means of numerous synapse terminals. (From Katz 1961, Copyright © by *Scientific American*.)

to the next. The impulse passing down a fiber as a wave of depolarization is an efficient system. This efficiency has been further increased among invertebrates by increasing the diameter of the fiber, for thick fibers conduct more rapidly. Such a strategy is found in the giant axon (nerve fiber) of the squid which runs down the length of the animal. Vertebrates have managed an even cleverer system to speed up conduction: they have encased the nerve in an insulating sheath made of fatty material (myelin) that has periodic gaps (nodes of Ranvier), and the wave of depolarization tends to jump from one gap to the next, thereby greatly increasing the speed of transmission.

The major evolutionary progression has been from a simple system of a few identical nerves, more or less equally spaced as in the nerve net in *Hydra* (Fig. 65), to a system with not only a great many more nerves, but with increased centralization of the nervous system. This is an evolutionary sequence in the same sense that there was an evolution of size and complexity. More primitive animals found in the earliest fossil-bearing strata have the simplest system, which becomes increasingly complex as one passes on to various worms, insects, and other arthropods, early chordates, fish, amphibians, reptiles, birds, and mammals. We take particular delight in this sequence because we place ourselves at the very top, and indeed homonids are the most recently evolved group of all those mentioned.

It is not especially important for us to examine all the anatomical changes that have occurred in the progression of the nervous system from *Hydra* to man, although something should be said about the size of the nervous system, or more specifically of brain size. This is an old subject, but one which is arousing renewed interest. Unfortunately the work has been done almost entirely on vertebrates, and part of the reason for this is that it is more difficult to estimate brain size in segmented invertebrates, which have conspicuous ganglia in each segment as well as an anterior brain.

The first point of importance is that there is a clear relation between body size and brain size; larger animals have larger brains. This can be shown on a logarithmic plot for different groups of vertebrates (Fig. 75). From this figure it is evident that brain size is strongly influenced by body size. We have said before that in the evolution of complexity, each level has a great range of sizes (Fig. 51), and therefore it is essential to separate that component of brain size that is independent of body size and truly reflects an evolution of complexity.

One can see in Figure 75 that fish and reptiles have relatively small brains compared to birds and mammals, and the brains of primates are larger relative to their body weight. Human beings do even better, along with dolphins, our only rivals. It is important to remember that these differences can, at least in the case of different primates, be explained in terms of heterochrony (Fig. 68). By changing the time at which brain growth stops relative to body growth, one can, in a rather straightforward way, produce relatively large or small brains.

There is another important aspect of the brain–body size relation. In Figure 75 are values for animals of widely divergent taxonomic groups, and the slopes of the points would fall on a line whose slope would be approximately 0.75 (that is, body weight $\propto$ brain weight$^{0.75}$). It has also been known for some time that if one compares more closely related animals, such as species within one genus, the slope of the line is less steep, and brain size increases only slightly with the increase in body

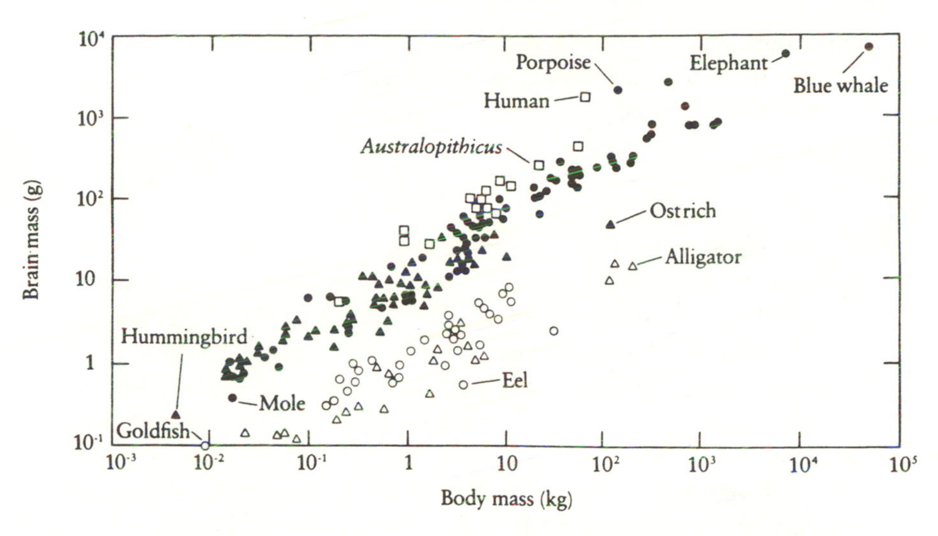

Fig. 75. The brain size of 200 species of vertebrates plotted against their body size on a log-log graph. Primates are open squares; other mammals are solid dots, birds are solid triangles, bony fish are open circles, and reptiles are open triangles. (From McMahon and Bonner 1983, Copyright © by *Scientific American Books*, redrawn from Jerison 1973.)

size (body weight $\propto$ brain weight$^{0.2-0.4}$). If one were to plot the degree of relatedness among animals against the degree to which the brain increases with body weight (that is, the range of exponents from 0.2 to 0.75), one could show that they have an inverse relation: increased relatedness means a decrease in the brain size relative to body size. The reason for this might be that in any small taxonomic group (for example, a genus) the time for the evolution of a divergence of the species has been short and the selection has been for change in body size; the brain has not had enough time to catch up. However, in unrelated, distant taxonomic groups there has been ample time to restore an efficient balance between the relative sizes of body and brain.

The lesson from these facts is that brain size and body size are not linked in any rigid way. Among related species there may be a selection for change in body size with relatively little effect on the brain. The only exception to this is the homonid line, where even in closely related species the brain increases markedly with body size. In other words, there may be a selection for body size changes relatively independent of brain size, and vice versa, in the homonid line. There have been numerous studies which suggest that there are a number of ecological factors, of which diet is an important one, that might also affect the brain–body size ratio.

Behavior

Unfortunately it is not clear exactly how brain size relates to behavior. We consider ourselves cleverer than all other animals and immediately assume this is so because our brains are comparatively larger, but the great brain size of dolphins and porpoises rivals ours and there is an enormous amount of speculation as to why this is so. Perhaps the answer lies in some middle course: brain size is in a very general way correlated with the complexity of behavior, but the correlation is imperfect, and we would like to know more about the exceptions. Here are a few facts that give one pause. It is often pointed out that there have been some men and women of extraordinary intelligence who have exceptionally small brains, Anatole France being a notable example. And we have all known people with magnificently large crania who seem to lack in quickness. This point is not too difficult to accept for one expects large individual variation, and we are all quite well aware of the fact that we have little understanding of the anatomical basis of human intelligence. Perhaps a better case for taking care in equating brain size alone with complex behavior comes from a view expressed by D. R. Griffin (1984). He points out the difficulty in knowing what goes on in the mind of an animal and shows that some animals very low in the evolutionary scale can perform exceedingly complex behaviors. In particular he chooses the honeybee, whose scouts can tell the other bees the direction and the distance of a source of nectar, and the bees that receive those signals can find the nectar with remarkable accuracy. We know that insects in general and bees in particular have excellent memories and an impressive ability to learn. We also know that a bee has very few neurons compared to even the smallest and most primitive vertebrate. Griffin asks, and the question is a good one, how do we know that the complexity of behavior is necessarily correlated with the number of neurons; how can we gauge what a bee thinks? The honest answer is that at the moment we cannot.

Despite these fascinating exceptions—the honeybees, the porpoises, and Anatole France—there is a general trend towards more complex and flexible behavior as one goes up the evolutionary scale from *Hydra* to man—from organisms of 11 cell types to organisms of over 120 cell types. What is unclear is whether this is strictly correlated with the number of neurons. Probably the principle that body size crudely correlates with complexity (Fig. 51) also applies to the nervous system. Remember that increasing complexity in animals means not only more cell types in general, but also more different kinds of neurons. There is an increase of nerve cells, both those with different morphology and those similar in appearance but with differences in the chemical signals they emit and the receptors they possess in their membranes. Certainly there is every opportunity for a high degree of complexity in a large central nervous system, but we still do not understand how it works in any profound way.

The next question is, what do we mean by simple or complex behavior? It is convenient to consider three main categories of behavior: one that is characteristic of single cells, one that is found within an individual multicellular organism, and one that is found between organisms, as in social groups.

Single-cell behavior. As we have already seen, there are individual cells that are receptors or effectors. Single cells can be both. This is most strikingly seen in motile bacteria, which can swim up (or down) a chemical gradient of some substance that either attracts or repels them. There has been much interesting recent work on this so-called chemotaxis of bacteria. It is known that there are receptors for the chemical on its surface, and if the cell is swimming towards an attractant, these receptors receive increasing amounts of the chemical from one moment to the next and as a result continue to swim. If, however, they happen to be swimming away from the attractant so that their receptors receive fewer molecules of the chemical at successive moments, then a signal is tranmsmitted to their flagella to reverse their motion momentarily. As a result the cell reorients before it resumes forward motion, and by repeatedly doing this as long as it is pointed in the wrong direction, it ultimately ends going up the gradient. These primitive cells clearly have both receptors and effectors and a means of transmission between the two. What is especially interesting is that the cell can compare the concentration of a chemical substance at two successive times; the ability to make this comparison has been termed "memory." In a strict sense this is correct, although no doubt the mechanism involved is totally different from memory in multicellular animals. The exciting aspect of the story is that much has been and is being done on the molecular basis of this behavior and it is already partially understood in important ways.

Also, it is well known from the early work of H. S. Jennings (1905) that some of the more elaborate protozoa show two kinds of important primitive behavior. He worked with the large ciliate *Stentor* and showed that if it was given a very slight stimulus at repeated intervals (carmine particles squirted gently towards it), at first the organism responded by contracting into a ball, but as the stimulus was successively repeated, the response became less and less until finally the stimulus was totally ignored. This phenomenon is known as "habituation." With a stronger stimulus (squirting the carmine more vigorously) Jennings found a different effect. After trying to avoid the jet, the *Stentor* would contract into a ball and then eventually swim away (Fig. 76). In this case the response is the reverse from habituation, for instead of ignoring it, it builds up for a large response, that of moving off. This behavior is called "sensitization."

Habituation and sensitization are thought to be general properties of cells and to play an important role in the neurons involved in synaptic connections of multicellular animals. If a nerve is repeatedly stimulated at a low level, its synaptic bulb will eventually show no response. On the other hand, if the recurrent stimulus is of greater magnitude, the synaptic bulb will fire off its chemical signal with sudden vigor. This is exactly parallel to habituation and sensitization in *Stentor*. There is considerable interest at the moment in trying to find a molecular basis for these behaviors, and some promising and plausible molecular models have been suggested.

Multicellular behavior. First it should be said that multicellular organisms, even the simplest ones, not surprisingly exhibit habituation and sensitization; these are very general, primitive behaviors. If we now look at the whole gamut of multicellular organisms, there is wide variety in the morphology and its increasing complexity in the evolution of animals. The question is: Are there behavioral correlations with that morphological progression?

Let us begin with the nerve net of *Hydra* and other cnidarians (Fig. 65). The main function of this net seems to be to coordinate movement when the contraction wave is transmitted by the network of nerves. Its importance to the proper functioning of *Hydra* can be seen clearly in the experiments of R. D. Campbell and his co-workers (1976) on individuals lacking nerve cells. (Campbell obtained these by treating the *Hydra* with a chemical that destroyed the i-cells, but T. Sugiyama and T. Fu-

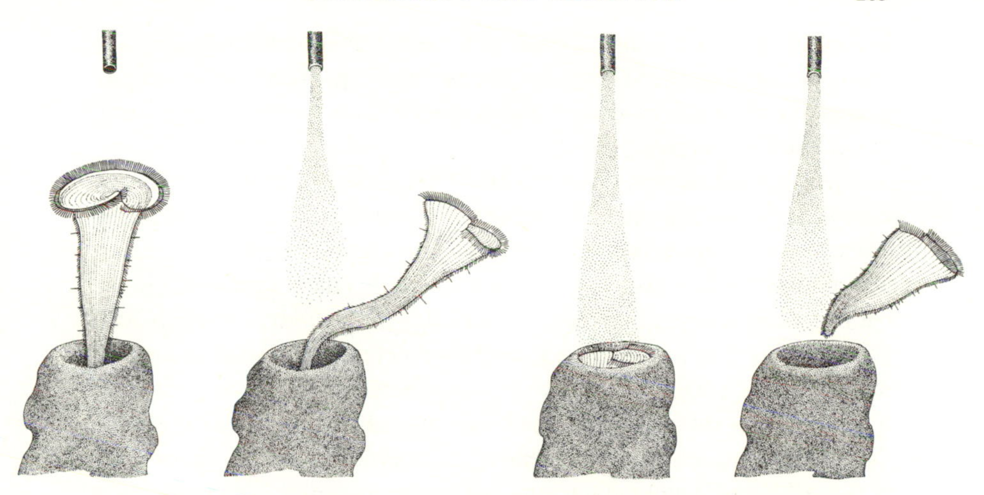

Fig. 76. H. S. Jennings' experiment on *Stentor*. A jet of carmine from a pipette is first avoided by the *Stentor*. Upon further disturbance the *Stentor* disappears into its sheath and ultimately swims away. (From Bonner 1980, Copyright © by Princeton University Press. Drawing by Margaret La Farge.)

jisawa [1978] produced similar nerveless *Hydra* by genetic mutation.) These *Hydra* had to be force-fed to keep them alive; they were incapable of wrapping their tentacles around their prey and engulfing them through their mouth. They also lacked the spontaneous activity of normal *Hydra*; they remained listless. They could be stimulated to contract, but the required stimulus was far greater than normal and the wave of contraction was exceedingly slow by comparison. Clearly the remaining cells have the basic ability to contract and pass on waves, but the nerve net is needed for more efficient contractions, and certainly for the coordinated movement required for feeding.

As one proceeds up the evolutionary scale, the next major step is the centralization of the nervous system, and with this appears an even more effective coordination of the movement. Various segmented animals such as annelid worms and arthropods illustrate this particularly well. They coordinate the muscle activities in a series of appendages which come off each successive segment. The sequence of movement of the segment appendages, be they swimmerettes or legs, is governed by the nerve pathways from one segmental ganglion to the next, in this way ensuring that the appendage of a segment can move only after the one just ahead of it has moved. However, these segmental nerve connections affect only the sequence, not the rate of movement. The latter is controlled in the brain; it gives off special fibers which go to each segment and give the signals for fast or slow. This dual centralization of anterior brain and ganglia in each segment allows remarkable control of complex movement.

These are not the only movements of animals with centralized nervous systems. For instance, by means of their photosensitive or chemical senses they can orient towards or away from light or chemicals, in this way finding food or avoiding enemies. To do this they must give different instructions to the muscles on either side of the body. This involves not only taking in information from light receptors or any other kind of receptors (such as chemical, sound, or electrical receptors), but also translating that information in the brain into the appropriate response in the correct muscles. In sum, a centralized nervous system has permitted both well-coordinated movements and an attuning of those movements to the circumstances of the immediate environment. Note that this is a great advance over the trial-and-error method of orientation found in bacterial chemotaxis; it heralds a new level of sophistication and organization of the nervous system.

This leads to the next major step in the evolution of behavior, which is the ability to learn, that is, to have some sort of systematic memory. Memory can be short-term or long-term, but the former does not include learning, for the message is quickly lost. On the other hand, it is uncertain whether any animal has just short-term memory alone. It seems rather to be the first stage towards long-term memory, and the two always go together. In all those organisms in which memory has been studied, especially in insects and in vertebrates, it is evident that one part of the brain is especially associated with short-term memory, and another with long-term. We hope that in the not too distant future we will understand the molecular activities in those two regions that would account for memory.

The learning and memory of insects can be quite remarkable, as has already been pointed out in the case of bees. They can remember many details about a landscape so that their flight to a source of nectar or back to the hive can be remembered and repeated. Some years ago T. C. Schneirla (1953) compared the learning performance of eight rats and eight ants in a maze of the same plan, and while the ants took three or four times more trials to learn the maze and be rewarded with food, they did learn it, making their ultimate performance almost as good as that of the rats.

Perhaps the main difference between the learning of invertebrates and of advanced vertebrates such as mammals is not just that mammals are faster at it, but that they can remember more of what they learn. It seems to be primarily a quantitative difference. Furthermore, if one compares the feats of human memory with those of an animal, there is again a big quantitative difference. Good police dogs or elephants are supposed to be able to handle in the order of a hundred different commands, and the learned vocabulary of chimpanzees and gorillas in captivity seems to be of the same magnitude. But consider an actor who memorizes the leading part of a play by Shakespeare, or Toscanini who memorized the entire scores of all the symphonies he conducted for a concert, or a friend's chemistry professor who knew the entire five-place logarithm table by heart. Admittedly it is difficult to explain the adaptive advantage of these particular examples in *Homo sapiens*, but presumably in our early ancestors there must have been some advantage that contributed to reproductive success in being able to remember many things. Clearly, then, one of the changes that has occurred in the course of the evolution of behavior is the increase in the capacity to remember.

Another important advance in the evolution of animal behavior is in the flexibility of response. Primitive invertebrates often have only one response to a stimulus—what is called a fixed action pattern. In contrast to this single action response, animals with more developed brains are often capable of making a choice of how to respond to a stimulus. I have called this multiple choice behavior, and it provides a way a stimulus can be assessed in a particular context so that the most appropriate of many possible responses may be chosen by the animal to the advantage of its well-being or safety.

Social behavior. As behavioral complexity increases, there is another phenomenon that appears. Higher mammals, especially humans, besides learning are capable of teaching. This presumably had its roots in imitation, as one animal in a social group, particularly a young animal, would imitate the others and in that fashion learn something that might conceivably be useful, such as a new way of obtaining food. This has progressed among the primates so that chimpanzees, for instance, may to some degree actively instruct their young. It is evident from various studies that chimpanzees not only teach the young by imitation, but the mother often reinforces the lesson either by encouragement, if the correct procedure is followed, or by the discouragement of mistakes. One of the characteristics of great apes is an extended period of parental care when opportunities for learning and teaching are prolonged. This is also one of the major differences between us and apes; the period of parental care becomes even longer, and the role of active teaching becomes exceedingly important. There is, as one might expect, a good correlation between brain size and this trend towards extended care of the young—an extended period of learning and teaching (Fig. 77).

Hand in hand with this increase in learning, and ultimately teaching, there is also a progression in the complexity of communication between animals. They can signal one another in specific or general ways which involve visual, tactile, and auditory signals, and even, as we have seen in the case of some fish, electrical ones. If we first look at invertebrates, we see that a wide range for this kind of behavior is not only dependent on the degree of complexity of the nervous system, but also on the extent to which the animal is social. One would not expect earthworms, which live in their burrows in the soil, to require much in the way of social signals. On the other hand, any social insect obviously would require a great many. Yet the nervous systems of the two allied groups, annelids and arthropods, are not enormously different.

It is interesting that, if one compares signaling among animals of different sorts, one finds that there is some specialization in the kind of signal used. Insects in general and social insects in particular rely heavily on chemical signals: trail substances, alarm substances, attack substances, nest recognition odors, and so forth. Tactile signals are also important, as often are auditory signals and to a lesser degree visual ones. Fish, in their aquatic existence, use the entire range: auditory, visual, electrical, and in some instances chemical signals. The mix varies also with amphibians, reptiles, and birds, and in mammals auditory and visual signals are common, while chemical signals still play an important role in many species. In the primates there has been an increased interest in the number of visual and auditory cues. For instance, apes have a wide variety of facial expressions, and, as has been known for a long time, these convey meaning to other individuals in a group. In this way they can show signs of fear, anger, pleasure, submission, and many other attitudes. It is also well known that many primates have a large variety of different calls or grunts, but only recently has some of the meaning been decoded. The African vervet monkey has separate alarm calls for different predators, so that the rest of the group will look up if

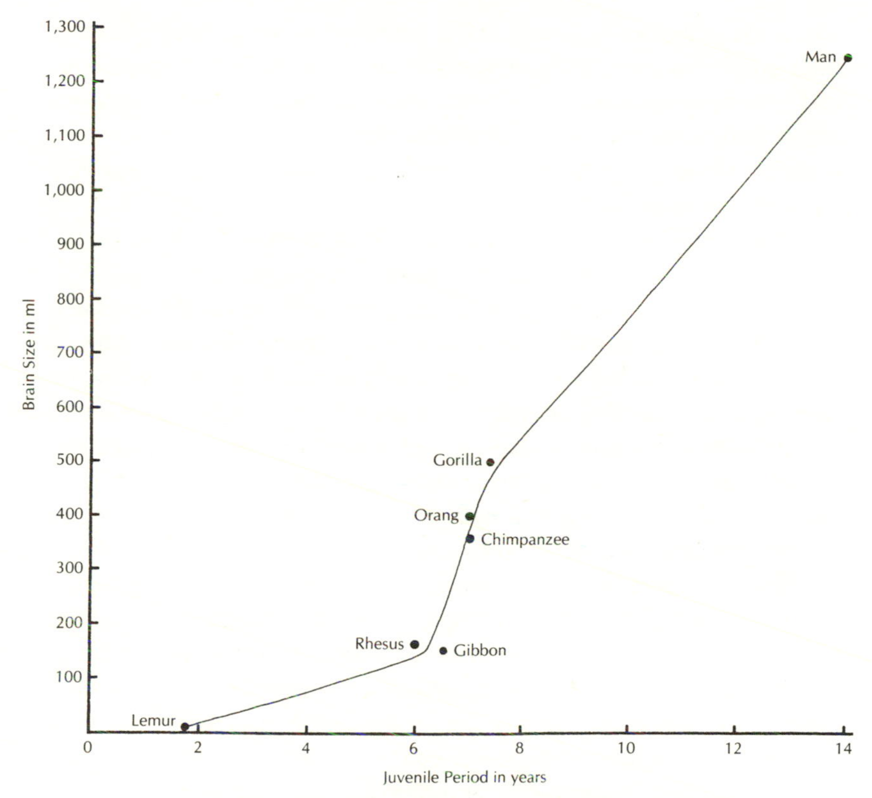

Fig. 77. The brain size of various primates and man plotted against the duration of the juvenile period. (Data on the brain size from Jerison 1973; data on the juvenile period from Napier and Napier 1967. From Bonner 1980, Copyright © by Princeton University Press.)

a hawk call is given and down for a snake alarm call and will know whether to freeze or flee. We manage far better with an enormous vocabulary which includes names not only for objects but for abstract concepts as well.

With this in mind we should be able to make some sort of graph showing the number of signals an animal can produce and plot it against the complexity of the nervous system. Unfortunately it would be impossible to construct, for not only is it difficult to count signals accurately but, as we have discussed, we have no precise gauge of brain complexity. It is certainly true that if one compares the most advanced social insect with any vertebrate, the vertebrate has many more signals.

The same trend can be found in the evolution of vertebrates. There is little doubt that there is an increase in signals from fish to mammals, and from most mammals to primates and then man, and within any one group from solitary to social. Unfortunately, until such measurements can be made properly these speculations seem idle; hopefully the future will bring us some hard facts.

There is another element of interest. Often there is a richness in the response that is not immediately evident in the signal alone. As W. J. Smith (1977) has pointed out in detail, signals are read in their context and, depending upon that context, can have different meanings. Among mammals a grunt may be little more than a signal for attention, and by looking at the signaling animal the others will know whether to flee or gravitate towards a new source of food or be on the alert for some potential danger. So much of the response depends on a combination of signals, for different mixes have different meanings. This combining of signals to make such distinctions, often very subtle ones, is again an increase in complexity. Furthermore, we assume that this is something achieved most effectively in complex nervous systems, and that the ability of bees or ants to make many of these fine distinctions in interpreting a set of signals must be comparatively limited. As the signaling system becomes more elaborate during the course of evolution, its very complexity generates even greater complexity by this method of producing nuances of meaning out of different combinations of signals.

Natural selection and behavior

We now come to a central issue concerning behavior and one which is crucial to my argument here. It is possible, through teaching and especially through learning, to pass on behavioral information from one individual to the next. This means that through behavior a new mode of inheritance has evolved, something quite different from genetic inheritance. It is a new unit of selection and therefore an invention of major magnitude, so let us examine it closely.

In the case of genetic inheritance it is essential that all the information be contained in one set of chromosomes. As was discussed at length earlier, the advantages gained in one generation in the form of a favorable mutation would be lost to the next unless there were some permanent way of encoding it in the chromosomes and passing it on to the offspring. This kind of inheritance has been strongly favored by natural selection, for it ensures continued stable reproductive success. Any change to any part of the body of a multicellular organism that is not passed on to the offspring obviously fails utterly, since it is selected out immediately. Now consider inheritance by passing some behavioral information from one individual to another. This kind of information has been called a "meme" by R. Dawkins (1976), and it is a useful term for it makes a suitable distinction from a gene. Memes may be any kind of behavioral information, any kind of idea, such as a custom, a way to avoid a predator, or even a fad. It is most helpful to keep its definition very general, thereby making it exactly equivalent to "behavioral information." The thing that sets memes apart is that they are learned by one individual from another, and they are not present in the DNA of the genes. Therefore, if an animal is removed from its mother before birth and raised in the absence of any of its own kind, it will lack all the memes that its family might have passed on, while it will retain all the genes that its parents bestowed upon it.

There are three main differences between genes and memes. One is that all genes come from DNA while all memes come from either teaching or imitation. The second is a consequence of the first: evolutionary change involving the change in frequency of a particular gene is exceedingly slow and requires many generations, while a change in a meme can occur in a matter of minutes or hours and does not even require a single generation. A meme could also be passed on from parents to offspring for many generations and persist for a long time, but nothing prevents it from disappearing instantly should there be a sudden alteration of the environment. The third difference is also a consequence of the first: genes can produce an organism that is capable of producing memes, but the reverse is meaningless. So memes are ultimately inventions of genes. It is true that memes can influence the direction of selection and in this way affect genes, but this is quite a different matter. Previously I have suggested that the brain and the genome have a symbiotic relation, that is, the brain can do things that the genome cannot, and for that reason it is an advantage to the genome to have an associated brain. In turn, the brain is utterly dependent on the genome for its existence.

It should be added parenthetically that in the last few years a large literature has emerged on how genes and memes interact. The ultimate aim in many cases is to try to dissect out how culture (defined as the transmission of memes) might interact with the genetic constitution of individuals to produce trends in cultural (i.e., nongenetic) evolution and in genetical evolution. Both are subject to selection, but some (including myself) like to preserve the term "natural selection" for the selection of genes, that is, for pure Darwinian selection. However, because the selection of genes has basically different dynamics from the selection of memes, the relationship between the two can be complex. In particular, the consequences of the dramatic differences in the rates of evolution by cultural versus genetic transmission can complicate or even invalidate a mixed dynamic model. It is not an easy subject, and the water has too often been muddied by the failure to keep the distinction between memes and genes clean and crystal sharp.

Plasticity. Earlier I took the position that an increase in plasticity means an increase in complexity, and it should be made clear why this is so. My definition of complexity for organisms was that complexity equals the number of cell types dividing the labor. In a situation in which, depending on the environmental circumstances, one can have various differentiations, there are increasing degrees of freedom in how the labor is to be divided. The manner in which complexity can be increased through plasticity is nowhere more evident than in behavior.

The first point that must be made is that the invention of behavior during the course of evolution itself opens up the possibility of increased plasticity. Let me give some examples which differ in their degree of obviousness. A desert mouse will stay deep in its burrow during the scorching daytime, but in the cool night it will come out to feed. If one area is devoid of food, any mobile animal can move on and search elsewhere. The most striking examples are birds that migrate and find rich sources of food in different environments the year around. In freezing weather we can put on warm clothes, build houses, and heat them with fires. In some ways this is very similar to the developmental plasticity we found in plants that respond in special ways to a variety of environments. In both cases there are alternative paths to cope with changes in the environment; in one it is achieved by switching developmental pathways, and in the other it is by responses in the behavior, usually of the adult or at least of a postembryonic individual.

An important point is hidden in the above examples. Some behaviors are rigid and invariable even though they may take alternative paths, depending on the circumstances. This is true of the plants that develop different leaf forms in or out of water (Fig. 65), and it is probably true of the desert mice mentioned above. Behavior can be rigidly programmed so that the response to the environment is automatic. The mouse probably does not say to itself, "My, it's hot this morning, I think I'll go below ground" (although it is true, as D. R. Griffin [1984] so correctly points out, we do not know what thoughts a mouse might or might not have). There are, however, many instances in which a response to an external signal seems automatic to a degree that makes it resemble a reflex more closely than a thought.

This leads directly to the fact that many behaviors seem to be rather rigidly programmed and occur in an invariant fashion. These are often referred to as instincts or innate behaviors, and it is presumed that varying amounts of the behavior pattern are directly encoded in the genes, just as the leaf shape alternatives of the crowfoot are known to be genetically determined (Fig. 65) to a considerable degree. For behavior it is totally unclear how the genes do this, but the evidence is overwhelming that some behavior patterns are strongly dependent on genetic make-up. First, let me give an example where this is so, and then I will give a second example where the same kind of behavior can be either directly inherited or, to varying degrees, learned.

These first examples are of songs, or auditory signals. In the cricket the sound is made by rubbing the wings together—there are rasp like structures where the wings meet, and as the wings move back and forth on one another the sprightly chirp of the cricket surges forth. The frequency of this so-called stridulation varies with the species; each species has its own particular code (Fig. 78). The male gives the call and the female is quite overcome and rushes in his direction. What she hears is also specific; she does not hear the code of related species but only her own. So the pattern of the frequency of the call notes is the same for both sexes, but one is a sender and the other is a receiver. To show that this is genetic, D. R. Bentley and R. R. Hoy cross-bred two species and found that the hybrids had a song that was intermediate between the two. Furthermore, this hybrid song was recognized by hybrid females who no longer responded well to the calls of either parental species. Nothing could be more clear-cut; the song pattern for the two sexes, both its production and its reception, is encoded in the genome.

The next example illustrates the point that song in some species of birds may be rigidly determined, as in crickets, while other species often have a large amount of learning; and there is reason to suspect that all of these permutations are adaptive and have arisen by natural selec-

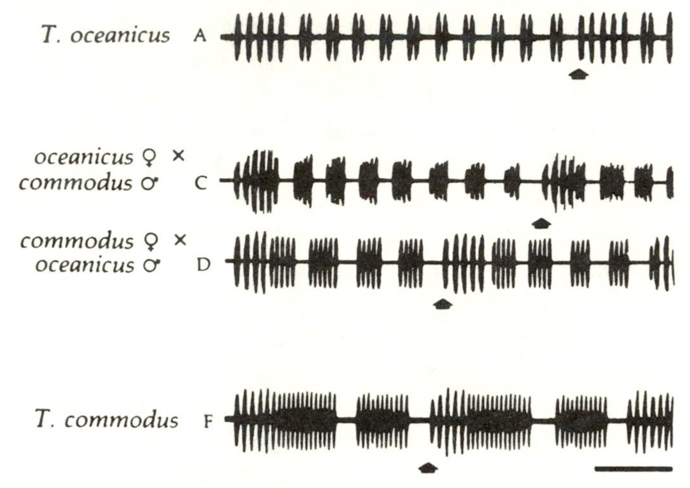

Fig. 78. Songs produced by two species of crickets and their hybrids. Each song phrase consists of a chirp and several trills. Chirps and trills are composed of pulses. New phrases begin at the arrows. The time marker at the bottom is 0.1 seconds. (After Bentley and Hoy 1972.)

tion. In the case of parasitic birds, such as the European cuckoo, the female lays an egg in the nest of some totally unrelated species, and therefore her offspring never sees its parents, but only its foster parents. If a female of the North American cowbird (a similar parasitic bird) is raised in isolation in the laboratory, as was done by A. P. King and M. J. West (1977), one can be absolutely certain it has never heard the song of the male. Once the bird has reached maturity, one can play a recording of a male and the female will immediately assume a copulatory position, even though this is the first time she has ever heard the song. Furthermore, the male's song is inherited, not learned. In this case the adaptive advantage is obvious: it is a way for birds that have never seen their own species to find one another and mate.

At the other extreme there are some birds that seem to inherit only a rough template of their basic song. Part of it is self-taught, as M. Konishi (1965) showed; if he destroyed the ear of a baby white-crowned sparrow, the deaf chicks had an even more rudimentary song. Normally, first with self-teaching and then by the imitation of others, the sparrows achieve an expert and finished song; there is a large component of learning. In such species one finds local dialects, and among the many studies of these dialects perhaps the most interesting is by P. F. Jenkins (1978) of the New Zealand saddleback, a grackle-like bird found on some isolated islands off the coast. By studying marked individuals on an entire island for five years, he was able to trace patterns of change in their song and the causes of these changes. This species is unusual in that song learning is not confined to juveniles, and adults are equally adept at imitating. The male will leave the home dialect group to enter a new one, and when it does it quickly adopts the dialect of its new community. In this case the advantage is presumably that if the new male is accepted, once he has adopted the song patterns of his new-found friends, this produces a desired degree of outbreeding, something that is clearly important in small populations.

Why dialects occur at all, as they do in many birds, is another interesting question. One possibility is that variation in song arose as a mechanism for success in sexual selection, which is the likely case in the New Zealand saddleback. Once the ability to imitate is acquired, it is not surprising that different populations of birds in different regions will have local dialects. If this hypothesis is correct, dialects are only a secondary byproduct of song learning for success in mating.

The next set of examples, also in birds, centers around predator avoidance. It is well known from early work that if newborn goslings are presented with a moving silhouette shaped like a hawk they will scurry for cover. There can be no learning in this, and its adaptive advantage is obvious: the goslings are prepared for danger the moment they emerge from the eggs, the period when they would be most susceptible to predation. But not all dealing with predators is best handled by having a fixed, innate behavior pattern. For instance, many birds mob a predator; they rush at a hawk or an owl, and by a noisy commotion they draw attention to the enemy who is thereby rendered powerless and cannot attack by surprise. The extent to which knowing what kind of an animal is a threat is learned, as has been demonstrated in a splendid experiment by E. Curio and his co-workers (1978). They put two cages, each containing a European blackbird, on the opposite sides of a narrow room, and between the cages they installed a box with four chambers at right angles to one another. By rotating the box 90°, the two birds in the opposite cages saw either an empty box or stuffed birds. At first, what the birds saw was not the same: one blackbird saw an owl, but the other saw a harmless Australian honeyeater. The latter was soon mobbed by the blackbird because it saw the other blackbird frantically mobbing. The bird seeing the owl was replaced with a naive blackbird, and the experiment was repeated in a way so that both birds saw the Australian honeyeater. The naive bird soon imitated the bird that learned to mob the harmless stuffed species and they both made a great commotion. This was repeated with a series of naive birds, so that the mob-the-honeyeater tradition was carried through six passages, one bird teaching the next. The obvious advantage to such a plastic behavior is that blackbirds, or any mobbing species, can quickly learn unfamiliar predators and do not require many generations of natural selection to be able to recognize a new danger. Another advantage is undoubtedly that, through memory, they can store more information and include all the different kinds of enemies they might have in one environment; behavioral memory can store more information than can easily be packed into the genes. Furthermore, it need store only the information that is relevant to a particular time and place to be adaptive. This is another way in which plasticity can be shown to produce increased complexity.

It is curious that in some of the examples given above the same behavior can be either directly inherited or partly learned. This implies that, at least to a limited extent, there can be, for any one behavior pattern, a switch during the course of evolution from a gene system to a meme system, or the reverse, apparently depending on the selection pressures. It is as though behavior has led the way for some morphological changes in evolution. A possible instance of this is in birdsong dialect, in which behavioral plasticity has produced regional dialects, and these serve an isolation function which might lead ultimately to morphological differences and separate species. Let me now give two other examples in which this appears to have happened.

The first example comes from sexual selection. In many species there is a great difference in the sexes: most often the male is either very much larger than the female, or more brightly colored. Fur-seal bulls may reach five hundred pounds, while the cows weigh only seventy pounds; and the peacock is quite splendid compared to the peahen. It was Darwin (1871) who pointed out that these differences might be caused by one of two different kinds of selection forces: a competition between males for the females, or choice on the part of the female for some special quality in the male. Let us consider the case of size differences between the sexes.

In seals in general there is a correlation between the size differences of the males and the females and the size of the harem, fur seals being the extreme case with the largest differences in size and the largest harems (R. D. Alexander et al. 1979). The way this size difference occurs is that female fur seals are fully grown and begin to breed when they are one year old, but the males cannot mate until they are at least seven years old; it is not until then that they are sufficiently large to compete successfully for a territory and a harem. There is an interesting instance of another mammal with a similar age difference in success in reproduction, yet there does not seem to be any appreciable difference in the size of the sexes. This has been found in a feral population of horses on an isolated island off the coast of North Carolina that has been extensively studied by my colleague D. I. Rubenstein (1982). He finds that the female first mates when she is three years old, but males cannot master a harem until they are about eight years old, even though they reach a mature size very early. It apparently takes that long for a male to be sufficiently canny and dominant in its competition with other males to succeed. Here behavior in the form of experience carries the day. Other large grazing herbivores, such as many deer and elephants, show size and other morphological differences between the sexes, although they are not so extreme as for fur seals. One could make a hypothetical case for behavior being the first way in which male competition is manifested, and that later it is reinforced by morphological change. The behavior might well not be innate, for one hardly expects experience to be programmed, but it has been replaced by a genetically determined size difference, if our conjecture of the sequence is correct. And if it is not, it remains interesting that there could be two such totally different ways of responding to the pressures of sexual selection.

The other example is in social insects. In primitive wasps the queen and the workers are approximately the same size and are indistinguishable in appearance. However, their behavior is radically different, for a queen is far more aggressive than the workers. And it is through her aggression that she manages so that all or most of the offspring are hers. If she sees a worker deposit an egg in a cell she will quickly rush in, grab the egg, and eat it, and then lay a substitute egg of her own. In wasps with more complex colonies and larger numbers of workers we see a size difference between the queen and the workers, and all signs of aggression seem to be absent. The smaller workers know their place and no longer try to compete for the royal prerogatives. This latter state is the usual one in all kinds of social insects. There seems to be harmony even among the different classes and sizes of workers and soldiers in ants and termites. The only aggression left in these more highly developed insect societies is at the time a new queen emerges, when tension will exist between her and the old one and the colony may split in two as a result. Again one wants to know why behavior has become transformed into morphological difference. The answer probably is that the morphological differences are a more effective way to divide the labor, because the tasks are most often best performed by differently sized or shaped workers. Queen attendance and care of the young are probably most effectively performed by the smallest workers, while only huge soldiers, often with fierce jaws, will be effective in defense. An extreme case is cited in G. F. Oster and E. O. Wilson (1978) of an ant species where the largest worker is thought to be the only member of the colony capable of crushing the large seeds which are an important component in their diet (Fig. 79). This is the ultimate in morphological specialization.

There is one final point to be made about these insect societies. It is very likely that the presence or absence of aggression is genetically determined, but in a way quite different from that in which the morpho-

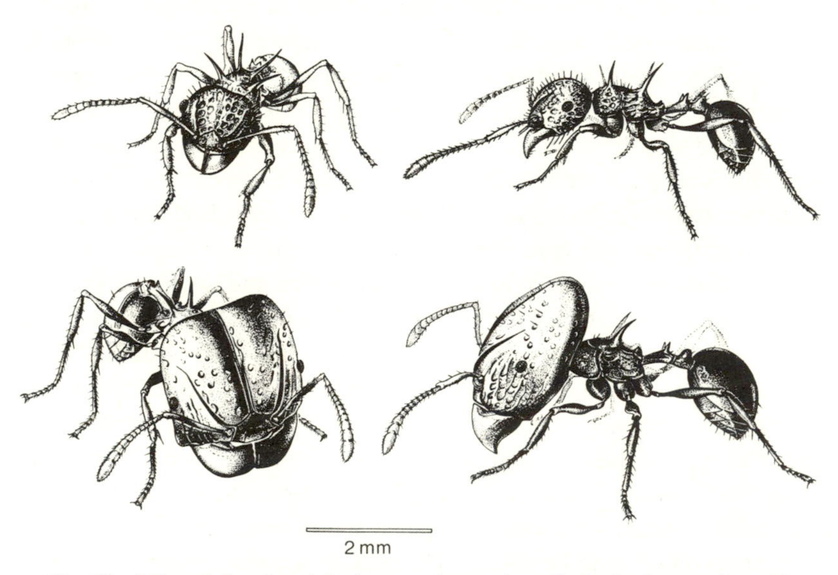

Fig. 79. Minor (*above*) and the huge major workers (*below*) of an ant found in the Celebes (*Acanthomytmex*). (From Oster and Wilson 1978, Copyright © by Princeton University Press. Drawing by Turid Hölldobler.)

logical differences are controlled. They are governed by a plastic response in the developing larvae or nymphs to differences in nutrition, pheromones, and other external factors. So in mammals we see a shift from relatively plastic behavior to genetically determined size differences, while in social insects the shift is from a rigid behavior pattern to a relatively plastic developmental one.

Another important feature of plasticity in behavior is the capability of making inventions. We think of much of human historical progress as one invention succeeding another, beginning with the discovery of the use of fire, tool making, and later the wheel, right up to the computer. This has vastly increased the complexity of division of labor within our society; it has deeply affected its entire structure. But lesser animals are also capable of behavioral inventions that can be passed on as memes, and there are some well-known examples: for instance, the tits in central Britain that learned how to peck open the milk bottles and take the cream at the top. This invention soon spread from one small area and now full milk bottles are not safe anywhere in the British Isles. As one might expect, primates are even better at inventing, and there have been many well-documented cases of a useful invention that then spreads in a population. The most cited example is of the Japanese macaque that lives on an island and devised ways of washing sand off potatoes and grains of wheat that were thrown on the beach. A clever female named Imo took the potatoes into the ocean to remove the sand, and later she threw the wheat kernels into the water and then skimmed off the wheat freed from the sinking sand. Both these inventions spread through the whole colony on the island and became an integral part of the behavior of all the monkeys.

Inventions are the pinnacle of achievements of behavior. Striking off in a new direction greatly increases the possible ways behavior may occur; inventions also are the ultimate in plasticity. It is as though there are no limits, at least on advanced complex behavior; it can invent its own new frontiers. No place is this better seen than in our own history.

Conclusion

The devising of the nervous system and the behavior that it can produce represent astounding advances in complexity. For individual organisms we defined complexity as a division of labor and used as our measure the number of different cell types. But labor is "function," and if one can have one cell type perform numerous different labors, the complexity is greatly increased.

Let me put the matter another way. It has been pointed out repeatedly that almost all the developmental steps are specified by genes, and that therefore the upper limit of the complexity that is directly controlled by genes is set by the number of genes. If one now considers all the information held by a social mammal—information that includes all its patterns of behavior, all its learning by memorization, and any innovations or inventions in its behavior that might occur within the lifetime of any one individual—then one can immediately see that a mammal contains far more total information than is found in its genome. The extra is a different kind of information, for it is stored in the brain and not in the nuclei of the cells in the whole body; but even more important, there is much more information than could possibly be stored in the genome. If it were all put in the genes, serious problems would arise. The genome itself would have to be so large that mitosis, meiosis, and all the mechanisms associated with sexual reproduction and cell activity might be impossible. A second problem would be that if all behavior were strictly encoded in the genes, flexibility would be lost. In fact, learning and memory would be useless and invention impossible; only genetic mutation could elicit change.

The nervous system and the behavior it produces has been an incredibly clever way of getting around this upper limit to genetic complexity. Not only that, but it is achieved within the embracing rules of the genetic system in a permissive arrangement that allows Darwinian selection by genetic transmission to take place alongside this new behavioral transmission of information.

Let us now review the way in which this has occurred. The first step, as we saw in Chapter 6, was to build the nervous system in a way that is quite different from other parts of the body. Besides producing a variety of different neurons by the standard division of labor among cells, the different neurons then form many interconnections during their early development. However, their ultimate pattern of interconnections seems to depend on function; use preserves particular connections, while those that are not used disappear and the cells die. Genes build the cell types and their ability to produce intercellular synaptic connections, and no doubt govern the initial number of cells, their initial placement, and their tendency to form initial interconnections. What interconnections remain as an integral part of the adult are those that are preserved by actual use. With such a method of development one can produce an enormous number of neuronal paths from the initial excess of neurons and synapses, few of which would have to be genetically specified. This is the first step in being able to store in an individual organism more information than its genes alone could handle.

The second step follows the first. The vast network of nerves that makes up an animal's brain is able to store information. Learning and memory result in an enormous stockpile of information, way beyond the storage capacity of the genome. This kind of behavioral storage has the further advantage of being able to change quickly to adapt to new situations and to be able to compete by cleverness. Therefore there has been enormous selection pressure for flexible behavior, and the result is a staggering complexity when we consider all the things we can accomplish with our own brains: the feats of memory, the power of communication in a large language, the ability to reason, to invent, to compose music, and the skill to concoct all the subtleties of thought of which we are capable.

Further References

The subject of the relation of brain size to body size has been discussed in many places. To mention a few, two recent reviews describe many of the aspects discussed in this chapter: P. H. Harvey and P. M. Bennett (1983) and J. S. Levinton (1986). (I would like to acknowledge my own indebtedness to my student Nadja Torres who helped me clarify my own understanding of the subject.) The matter of bacterial chemotaxis is well reviewed in Alberts et al. (1983). Molecular models for sensitization and habituation have been developed by E. R. Kandel (1976). I discuss multiple choice behavior in Bonner (1980). Teaching in apes is described briefly in F. E. Poirier (1977). The recent work on signals with specific meanings in vervet monkeys has been done by R. M. Seyfarth and D. L. Cheney (1984). There is a large recent literature on the relation of gene and meme inheritance. It is not an easy subject, but for the brave who wish to pursue it, here are a few references (and they cite others): R. D. Alexander (1979), L. L. Cavalli-Sforza and M. W. Feldman (1978), W. H. Durham (1982), C. J. Lumsden and E. O. Wilson (1981), and P. J. Richerson and R. Boyd (1985). The work on the genetics of cricket song and other topics in animal behavior may be found in J. L. Gould (1982).

Chapter 8

The Evolution of Complexity: A Conclusion with Three Insights

Summary of the argument ▪ *The somatic vs. genomic complexity insight* ▪ *The size-complexity insight* ▪ *The integration-isolation insight* ▪ *Summary*

Summary of the Argument

Thus far in this book I have developed the following arguments. There has been, during the course of evolution, an increase in the upper limit of the size of animals and plants, and this can be accounted for by natural selection. A selection for size increase has meant new worlds to conquer, either to avoid predation or competition or to be especially successful as a predator or competitor. This does not mean that small or intermediate size organisms have disappeared; to the contrary, every biological community on the globe today contains a complete spectrum from the smallest to the largest organisms. Furthermore, there is an inverse relation between size and abundance: the larger the animal or plant, the more scarce it will be in any one environment.

There has also been an extension of the upper limit of complexity during the course of evolution. This is evident on two levels. Within organisms it has meant first a great proliferation of the number of cell types; there has been an increase in the amount of differentiation during development. In animals it was possible to extend the upper limits of complexity by first inventing a novel way of constructing a nervous system, and then having such a system capable of behavior, thereby greatly adding to the possibilities for complexity. On the second level there has also been an increase in the complexity of animal and plant communities, that is, there has been an increase in the number of species over geological time, and this has meant an increase in species diversity in any one community.

If one examines the relation between size and complexity of living organisms, one can make the very broad generalization that larger organisms are more complex (i.e., have more cell types) than smaller ones. However, the correlation is extremely rough. It is true that an increase in size mechanically requires a certain degree of division of labor, but clearly natural selection must be acting independently, within certain mechanical limits, on complexity and on size. In some environments there might be a selection pressure for size increase or decrease that would in no way affect the number of cell types; in others there might be a selection pressure for an increase or a decrease in complexity without any change in size. These changes might involve a reduction in the number of cell types due to the adoption of a parasitic existence, or they might involve an increase in the number of cell types favoring greater efficiency in the functioning of the phenotype with a corresponding increase in success in reproduction. It is quite possible that selection for size change and complexity change could be occurring simultaneously, and the two might or might not be the result of the same selection pressures.

In this final chapter I would like to show that from these considerations of size and complexity one can come to some major insights which say something important about the way evolution has occurred during the history of the earth.

The somatic vs. genomic complexity insight

First, let us reexamine the hierarchical levels with which we have been concerned. We have stressed two levels: within an individual multicellular organism and between organisms in a community. It will be immediately obvious that as useful as these categories have been, they are imperfect, for there are many examples we have given which do not clearly fit into either one. For instance, the distinction between populations of the same species and populations of different species are both in the "between organisms" level, yet they are very different. There is also the problem of social animals, expecially social insects, that seem to be in both levels: they are made up of separate multicellular individuals, yet they develop in a manner closely resembling that of an individual organism.

These difficulties may be circumvented if we consider another way of identifying the levels. The individual may be considered a level within which differentiation or internal diversity is nongenetic. The genes within all the cells of a differentiated animal or plant (with some esoteric exceptions which need not concern us here) are identical; all the differences are reflected in the cytoplasm. We could, therefore, call this a level of *somatic complexity*. Such a label would include the nongenetic complexity found in behavior. Furthermore, it would apply to animal societies where the morphs or castes found in social ants or bees or naked mole rats, or the behavioral differences in the individuals in a group of social birds or mammals, are generally nongenetically determined or somatic in nature.

(There is a technical point here which deserves an aside. It is quite possible that some of the behavioral differences among the animals in a social group might have a genetic component. It is conceivable that the position of an animal in the dominance hierarchy or pecking order of a group might to some degree be affected by its genetic constitution. However, it is well known that external factors such as starvation or disease will cause a dominant individual to become subordinate. The genetic differences that might occur between individuals never determine the entire pecking order of a social group but could affect only the degree of aggressiveness of individuals, and the ultimate position of those individuals will depend on their interaction with fellow members of the group and on other external conditions.)

Notice that by referring to this level of "within organisms" as somatic complexity, it is the division of labor that is stressed, as is the fact that the division of labor occurs among genetically identical cells or individuals. As a result it is possible to include behavior and social animals. Because genetic differences are unimportant to this kind of division of labor, it obviously all occurs "within a species." Such a label is yet another way we could identify this hierarchical level, in contrast to the next, which would be "between species."

Earlier we called the second level "between individual organisms," but now we are adding that these organisms are not genetically similar but separate species. Again this puts the emphasis on diversity or the division of labor; we are not talking about populations of the same species. If we call the first level one of somatic complexity, we can, by contrast, call the second level one of *genomic complexity*. It is important to emphasize that at this level the diversity can be directly controlled by natural selection, for it involves the encouragement within a population of those genes that favor reproductive success and eliminates automatically those that do not. In the case of somatic complexity it is only the *ability to vary* (within a certain range) that is genetically determined; the variation which appears is controlled by external or environmental factors, and therefore the genetic effect is weak and remote. Such is strictly not the case with genomic complexity, where the variation is under direct gene control.

It is also important to note that by placing emphasis on how variation is controlled differently in the two levels, one is making the point that they represent two sides of the coin with respect to sexual reproduction. At the level of genomic complexity, all the individuals (with exceptions such as organisms that have asexual life cycles interspersed between periodic sexual ones) are sexual, either male or female, and reproduce by mating and producing a fertilized egg that is the next generation. At the level of somatic variation, that fertilized egg becomes diverse in its own, nonsexual way. All the genes of all the cells in the embryo, or of all the cells in the brain, are genetically identical (the known exceptions are trivial), and therefore all the differences that arise are extra-genomic, either in the cytoplasm or in the behavior of the nervous system. Presently, in the discussion of the third insight, I will examine the relation of sex to the evolution of size and complexity in more detail.

To conclude, let me state the somatic versus genomic complexity insight in a succinct way. Biological complexity or diversity comes in two forms: one in which the genomes are identical (or very similar) and any variation is phenotypic or somatic, and the other where the variation is in the genomes themselves. Ultimately both kinds of variation are controlled by genes, but somatic complexity is further removed from the direct gene instructions and more influenced by external factors in the immediate environment.

THE SIZE-COMPLEXITY INSIGHT

In Chapters 5 and 6 it was shown that size correlates in different ways with internal complexity (cell differentiation or somatic diversity) and external complexity (species or genomic diversity). In the former, as animals and plants become larger, there is a general trend towards an increase in the internal division of labor, or the number of cell types. It is true, as is evident from Figure 51, that the trend is a gross approximation because for any one level of complexity the range of sizes of organisms is very large. But the generalization holds if one considers the largest member of each level of complexity or the smallest. It is best to imagine that there must be a small range of especially efficient sizes for a particular complexity level and that subsequently natural selection caused some organisms of that level to become larger or smaller, in this way creating the broad range. This hypothetical, efficient range of sizes would increase directly with the internal complexity of the organism.

In contrast, species diversity seems to decrease, rather than increase, with the size of the organism, at least over the range of larger sizes for any one group of organisms, as is shown in Figures 42 to 46. It is not clear why this relation abruptly reverses itself for the very small animals and plants, although there are a number of possible explanations, more than one of which might apply. Possibly for larger organisms, such as vertebrates and higher plants, there is an optimum size, and below that size there are complications in the construction of the body and therefore fewer species. In microorganisms the reason is more uncertain. In any event, we will make the assumption that the reason for this abrupt drop at the lower end of the curve is a secondary phenomenon, and that our concern should be with the main portion of the curve which shows clearly that the larger the organism within any one group, the fewer the number of species for that size.

We can now represent these two trends in a very simple, nonquantitative diagram (Fig. 80). Both internal complexity in the form of the number of cell types and external complexity in the form of species diversity are plotted against the size of organisms, and from this it is obvious that as the internal division of labor increases, the external diversity decreases.

This diagram (Fig. 80) illustrates the size-complexity rule. Let me now state it in words. In any one portion of space in nature, the larger the size group of organisms, the fewer the number of species in that group. But as species diversity decreases with an increase in the size of the organisms, their internal diversity, or differentiation into cell types, increases. Therefore, in a rough way the total amount of diversity remains similar for different size groups, but as one goes from small to large organisms, the external (genomic) diversity goes down, and the internal (somatic) diversity goes up.

It is helpful to think of this relation in the concrete terms of an example. Earlier we described a temperate forest in which the largest animals were mammals of roughly a meter in length, and then we descended the size scale by an order of magnitude of the linear

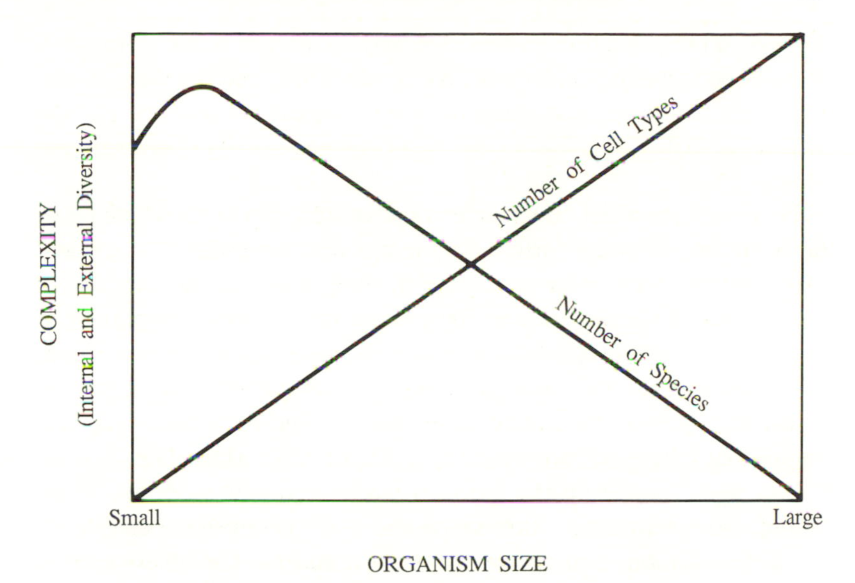

Fig. 80. A diagram showing the relation between internal diversity (number of cell types) and external diversity (number of species) for different-sized organisms.

dimensions, coming next to numerous birds and newts, frogs and toads, and small mammals which are in the 10-centimeter range. The next group, in the 1-centimeter range, would include various insects and worms. In the 1-millimeter range one finds a great variety of invertebrates including insects, nematodes, and mites. Organisms a tenth of a millimeter long (100 mμ) would comprise a wide group of rotifer worms, various protozoa including ciliates, flagellates, and soil amoebae. Finally, the organisms 10 mμ and smaller would include the vast horde of different kinds of bacteria. One can do a similar story for plants from large trees to small bushes, grasses, fungi, simple molds, small algae, and finally bacteria.

If one examines such series, it is quite obvious that there are fewer species of large mammals than of small mammals, birds, amphibians, or reptiles; and in turn there are many more species of insects and worms. The difficulty comes in the realm of microbes because of our uncertainty as to what constitutes a species in those minute forms. At the same time it is obvious that in a general way larger organisms, that is, vertebrates, have more cell types than the invertebrates. The reduction in the number of cell types continues down through the smaller rotifers and finally to minute amoebae and bacteria. It is indeed true that these trends must be depicted with a broad brush, and the same is true for plants, but there is no question that the trends are real and that the size-complexity rule holds for any particular community of animals and plants.

Let us now consider how this relation would apply to the whole range of evolution, from the earliest beginnings to the present. It is evident that there has been, over the course of time, a continuous increase in the number of species, and an increase in the size and complexity of organisms. Therefore, all three of the parameters we have in Figure 80—size, internal and external diversity—will all increase during the course of evolution. By adding the element of time to the basic diagram, we see that it becomes progressively enlarged (Fig. 81).

This again brings up the question we have raised more than once before. Does natural selection act on one of the parameters (size, internal differentiation, and species diversity), and are the others carried along for purely mechanical reasons, or can selection act independently on each? The answer is that selection acts on all three, but it remains true that if one is altered it may automatically affect the others. Let us briefly review the case for the three separately.

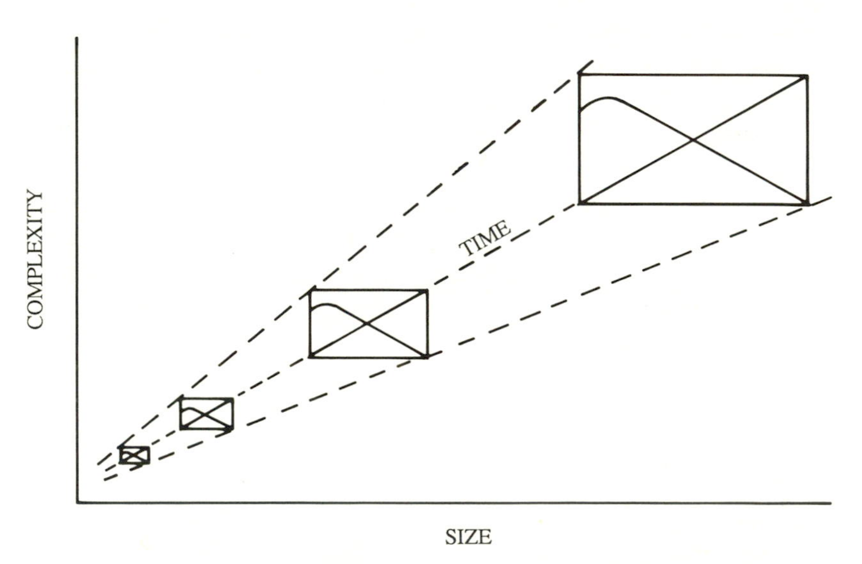

Fig. 81. A three-dimensional graph showing how internal and external complexity and size increase with time during the course of evolution.

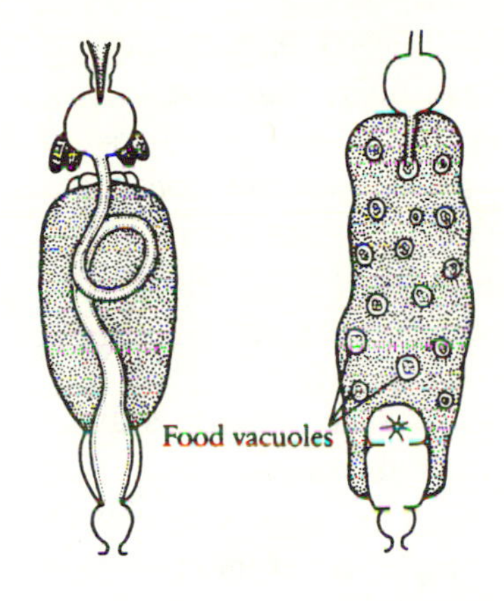

Fig. 82. The conventional gut system of a typical rotifer (*left*) and the acellular gut of an unusually small rotifer (*right*), showing food vacuoles passing through a continuous cytoplasm. (From McMahon and Bonner 1983, Copyright © by *Scientific American Books*.)

1. *Size*. As is amply evident, there can be a selection for size. If the selection is for size increase, there may well be an upper limit set by the number of cell types. Or if there is a selection for size decrease, there again may be limits imposed on the size of certain parts. For instance, it was pointed out that the hearts of hummingbirds are relatively large, presumably because if they remained proportional to body size they would have to beat at an impossibly fast rate. In some instances a selection for size decrease may result in a loss of cell types, as in parasites or in rotifer worms, where with size reduction the cellular gut tube disappears in some species and is replaced by a protozoa-like continuous protoplasm where the food is taken into small vacuoles (Fig. 82). From this we can conclude that there may be selection not only for overall size changes but also for changes in the relative sizes of parts within an organism.

Size changes, on the other hand, have a direct effect on species diversity because a common way of forming two species from one is to have simultaneous selection for a large and a small form, often specializing on different-sized foods. This is thought to be the case for many birds such as the New Guinea pigeons described by J. M. Diamond (Fig. 21), or some species of Darwin's finches (Fig. 20).

2. *Internal differentiation*. There are many reasons for believing that there is a direct selection for an increase in internal differentiation or somatic complexity. Because each major group of higher animals and plants arose from a small ancestor, one suspects the subsequent size increase was made possible by the advantage gained from increased

efficiency. This is the most important way in which natural selection can change the individual organism. Within limits, size can change without any change in the internal division of labor, and the number of cell types can change independently of any size change. But if one looks at the wide range of sizes for different degrees of internal complexity, it is clear that the most important advances involve complexity, although they do correlate with size.

3. *Species diversity.* Changes in species diversity are also a major activity of natural selection. The whole matter of how one species may turn into two is a central subject in population genetics and evolutionary biology. In some way a population becomes split, either into two separate geographic regions or as two populations living together which become isolated in their reproduction and do not crossbreed. In small populations the splitting may occur also by random changes (genetic drift). In any case, the populations become sufficiently different genetically so that they evolve into separate species. The differences may be in size, but can equally well be in shape or, for some animals, in behavior pattern. In general, this external diversity does not seem to be directly related to the internal diversity of differentiation, although in those instances where a new cell type emerged during the course of evolution it is presumed that the change might have been one that was also responsible for the formation of a new species. However, such cases must have been rare; more often the change is a less fundamental one. In any event, there are opportunities for new species as new niches in the environment develop. In some cases those species-forming steps might involve increased efficiency of some crucial sort, perhaps affecting feeding, or the catching of prey, or the avoidance of predators.

From this discussion one can conclude that evolution usually progresses by increases in complexity. This means an increase in the number of cell types within individual organisms, which in turn has increased the efficiency of the multicellular body, be it that of an animal or a plant. In terms of a community of organisms it means that new species have formed; there has been an increase in diversity. The traits that have been responsible for the evolution of one species into two have been manifold and include differences in internal structure, including the possibility of an increase in structural complexity, in behavioral complexity (in the case of animals), and of size. Therefore, even though natural selection acts on both size and complexity, it is clear that complexity is the more significant, the more central of the two. This is so in the sense that a new level of complexity opens up a whole range of sizes which overlap with those of the previous level and extend upward to previously unattainable sizes. Conversely, a small increase in size is always possible at the upper range of sizes for a given level of complexity, but at the cost of lowered efficiency and thus at a disadvantage in competition or predation with better-adapted beasts of similar size but at the next level of complexity. In this way one could account for reptiles being supplanted by mammals, and ferns (pteridophytes) being supplanted by gymnosperms and then angiosperms (Fig. 83). The next insight I shall discuss is one which follows directly from what has just been said and involves complexity alone.

THE INTEGRATION-ISOLATION INSIGHT

It is an interesting paradox that at all levels of complexity there seem to be forces which bring components together, as in the integration of cells into individual organisms, or gatherings of organisms into social groups; but at the same time there are other forces which work in the opposite direction and cause those individual organisms or those social groups to be isolated from one another. Curiously, the forces which seem to be working in opposite directions producing integration and isolation are the same; they are the forces of natural selection. First, let me show in a general way why this is true, and then I will give detailed examples for the hierarchical levels.

The reasons why natural selection would favor integration are quite obvious. By bringing smaller components into integrated larger units, it is to be expected that greater efficiency would result, and such efficiency would be encouraged by selection, that is, it would increase reproductive success. Therefore, the gathering of cells into discrete organs, or the gathering of cells and organs into discrete multicellular organisms could be accounted for entirely on the basis of efficiency. The same would certainly hold true for social groups of animals; there are many arguments which can be raised to show that social groups gather food or protect themselves more effectively than solitary individuals, and selection pressure for the integration of social groups is the result of individuals in such groups being more successful in reproduction.

There are two ways in which isolation is promoted by natural selec-

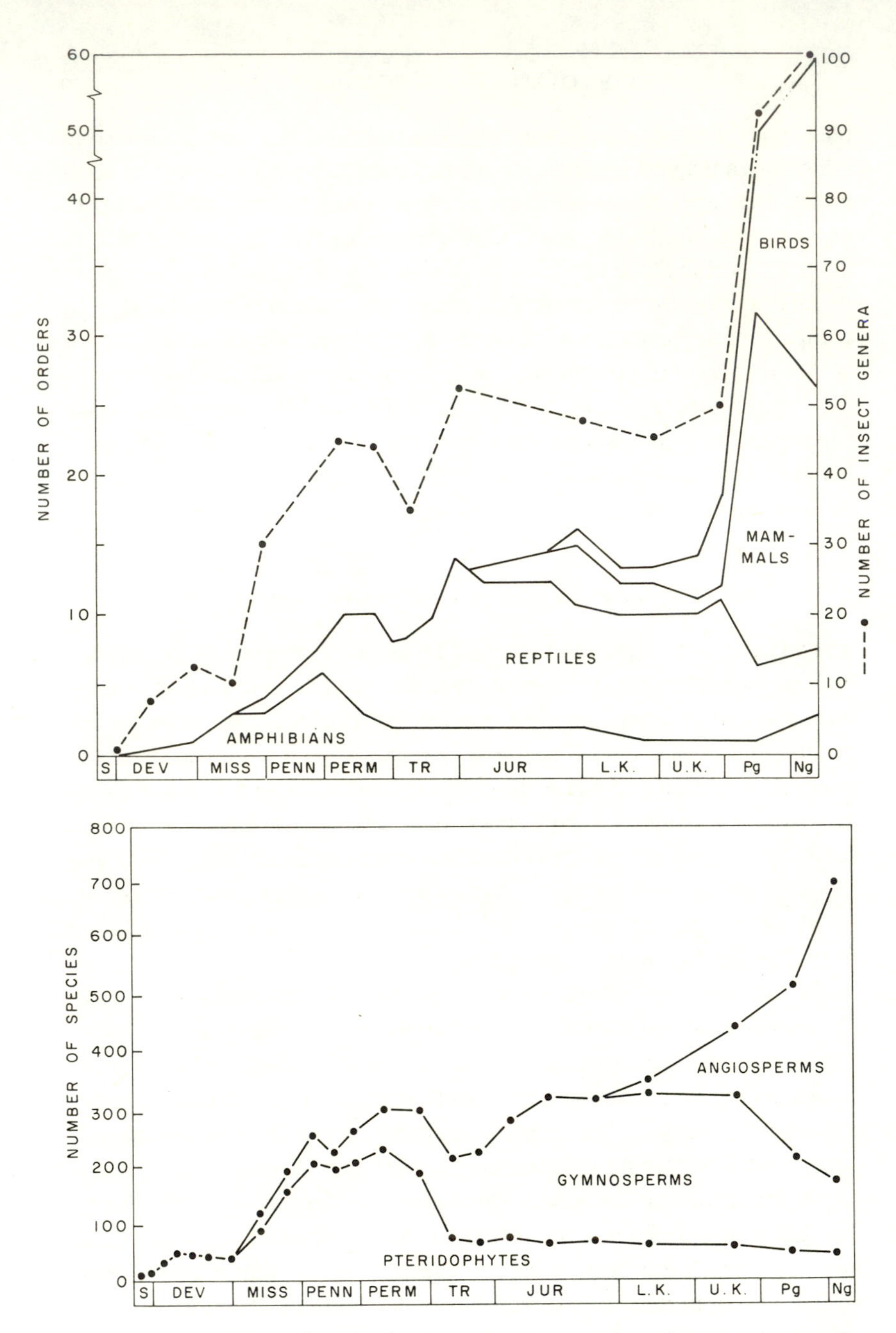

Fig. 83. The change in diversity for animals (*above*) and plants (*below*) over the last 400 million years. Note that not only is there an overall increase, but sometimes with time one group will dominate over another, e.g., mammals over reptiles and angiosperms over gymnosperms. (From Niklas, in *Patterns and Process in the History of Life*, Raup and Jablonski, eds., 1986.)

tion. One is by the production of units of optimum size. If integration were allowed to continue without check, one would find larger and larger units—so large that they would soon become inefficient and therefore be at a selective disadvantage. The other, more important reason for the prevalance of isolation has to do with selection pressure for controlling the amount of variation towards some sort of an optimum. The argument is straightforward. There will be a constant selection pressure for increased variation, for it is only by producing variants that organisms can successfully perpetuate themselves. But too much variation will be selected against because new, successful variants will be lost by excessive change. The amount of variation can be controlled by governing the mutation rate in asexual organisms, although much greater control can be obtained through sexual reproduction, with its clever device of meiosis followed by fertilization. Once invented, there would clearly be strong selection to retain the sexual system, and such a system involves the isolation into separate sexes (and integration at fertilization).

Having asked why we have the simultaneous existence of integration and isolation, we may now turn to the question of how they are achieved mechanically. It is remarkable that at all levels they are carried out by means of signal-response systems which can operate between cells or between individual organisms.

Somatic complexity level: Integration. Perhaps the most striking example of an integrating mechanism, which involves a signal-response system, may be found in the aggregation stage of the cellular slime molds (Fig. 84). Here all the amoebae are capable of giving off a chemical signal or attractant (called an acrasin) which manages to guide the cells to central collection points by chemotaxis. It is thought that after a period of starvation certain cells begin to secrete the acrasin sooner than the others, and those cells become the foci by attracting the cells immediately around them. The acrasin is given off by the central cells in a pulsatile fashion. As a wave of high concentration of the acrasin passes outward, it first attracts the cells that it passes over, and then induces them to secrete their own acrasin, which passes along to attract the next group of peripheral cells in the same fashion. The center of the aggregate sends off these pulses of acrasin at the highest frequency; it is the pacemaker, and all the cells become subservient to it. In the most direct sense this aggregation is an act of integration. The acrasin is a

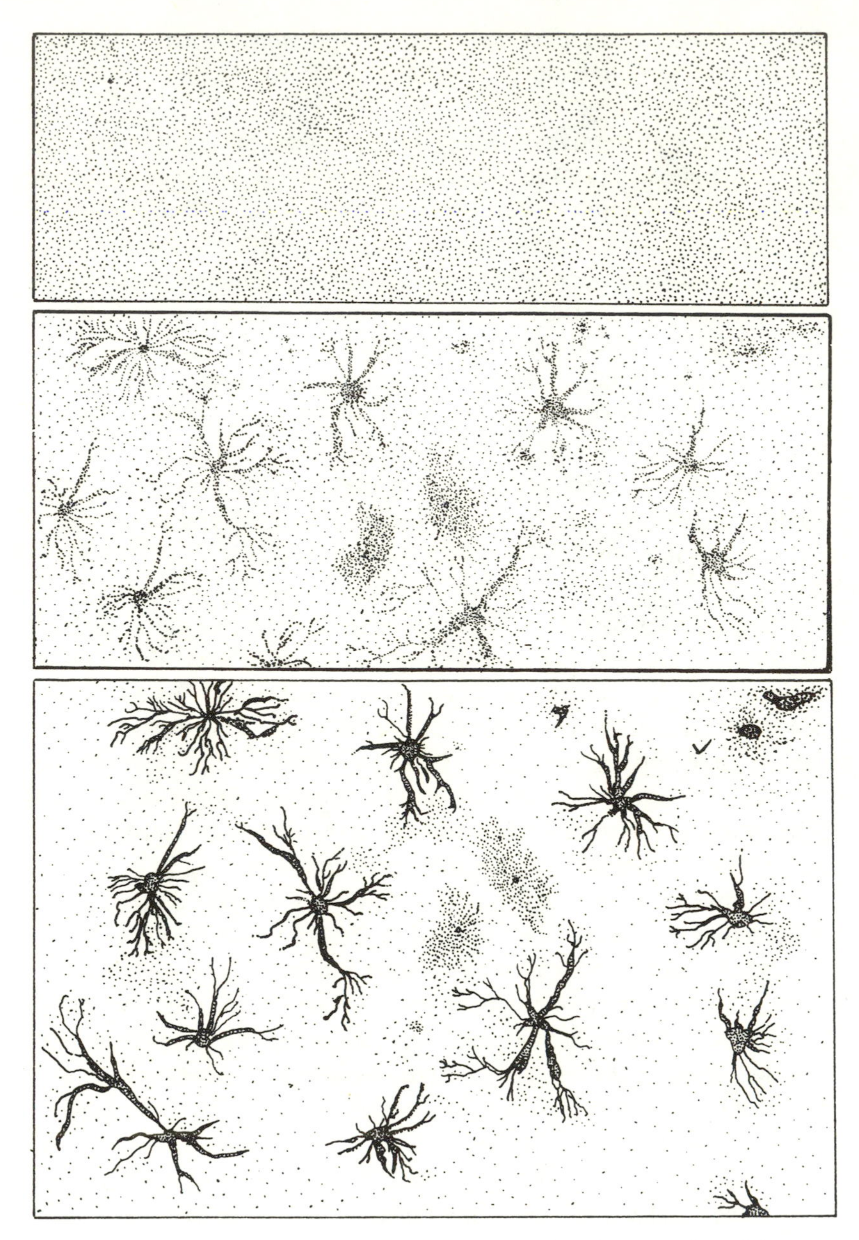

Fig. 84. Three different stages (semidiagrammatic) of aggregation of the cellular slime mold *Dictyostelium*. (From Bonner 1974, Copyright © by Harvard University Press.)

small molecule which diffuses readily, and the responding cells have specific receptors for it. In this respect it is no different from vertebrate nervous and hormonal systems.

The aggregation by acrasin is by no means the only integrating signal-response system in the cellular slime molds. Others are found in the later stages of development; for instance, there is evidence for the existence of signals involved in the formation of spores and stalk cells, and somehow those signals manage to see that the ratio of stalk cells to spores remains proportional regardless of the size of the cell mass.

Integrating signal-response systems are abundant in all multicellular organisms, the majority of which are not aggregative but develop from a fertilized egg. As an obvious example, embryonic induction is a signal-response system that integrates, or helps to organize the internal construction of the embryo. In plant development all the growth hormones serve the same function, for they manage to produce a proportionate and harmoniously constructed plant. In fact, it is difficult to conceive of any step in development that does not in some way involve an integrating chemical signal-response system.

Another system of this sort, common to all plants and animals, is found in the process of fertilization. It has been known for many years that there is a chemical communication between egg and sperm, and that the penetration of the sperm is not possible unless it gives off the proper chemicals, usually on contact with the egg; and the egg in turn must have the appropriate receptor system, which may be quite elaborate. In some species the egg gives off a chemical which can orient the correct sperm chemotactically so that it will swim towards the egg. In this case it is the sperm that has the specific receptors so that it can respond in an appropriately oriented manner.

Some of the most interesting cases of integration are found in social animals and involve behavior. Animal societies are enormously diverse in their structure, and in order to obtain an excellent panoramic review the reader is urged to consult the well-known book of E. O. Wilson (1975). Here I will pick two examples, one from the social insects and the other from mammals, to illustrate how animals can become integrated in a group.

In social insects there are two ways in which integration takes place. One is during the continued development of the colony. First of all there is an elaborate division of labor controlled so that there are the correct proportions of different size workers. Also, because the colony has a continuing development, if some class of worker is accidentally reduced, the original proportions will be restored by the further emergence of young individuals. As was pointed out earlier, this control over the division of labor is carried out by means of a number of external factors, the most important of which are nutrition and pheromones. Both are chemical signals which stimulate the appropriate responses. Pheromones trigger the internal hormones (ecdysone and the juvenile hormone), which in turn affect the organs within the body of each individual so that it develops into a soldier or some other form, depending on the signals.

The other component for integration within the colony involves behavior. It may be that insects' behavior patterns are mostly rigidly programmed; nevertheless, they involve an interchange of signals between individuals that have immediate effects (as compared to slow, developmental ones). In ants, for instance, if an individual has found food it will lead the other workers to the source by emitting a trail substance on the direct path, and the others will follow. Or those who discover there is a danger will give off an alarm substance which throws all the others into a great frenzy; and, according to their size, they either go out to attack the enemy or rush towards the nursery and carry the larvae and the pupae to a safer place. In other instances the signal is tactile; for example, ants that are replete with food can be persuaded to regurgitate some droplets if they are stroked in the correct fashion.

It is important to remember in this example that behavior uses the same signal-response system we found in the nervous system or, for that matter, in the communication between any of the cells within the body. The only differences are that the signals are not all chemical but may also involve pressure, hearing, and sight, and that the signal generally has to span a considerable amount of space between individuals (touch being the exception).

For comparison, let us examine a mammalian social group such as a wolf pack. The big difference between it and the insect society is the degree of plasticity in the behavior. There is less physical difference between the members of the pack—nothing comparable to the great variety of insect workers (although a fascinating exception among mammals is the African mole rat, where one finds castes within the group; see J.U.M. Jarvis 1981). In wolves there is the difference between the sexes and between the adults and the young. Among the adults there are more than one male and female in a pack; there may be from four to almost twenty in all. A strict dominance hierarchy is set up so that there is an alpha male and an alpha female, the king and queen of the group. This hierarchy is established entirely by behavior, by a jockeying for the top position among the individuals in the group. B. J. Cole (1981) has shown that such dominance hierarchies also occur in some ants, which suggests that the difference between insect and mammal societies is not so great as one might have supposed.

There are many ways in which the wolves in a group communicate with one another to cooperate effectively in rearing young and capturing food. Those that return from the hunt are welcomed with a greeting ceremony involving great tail wagging and friendly posturing, and as a result the food will be shared by the hunters with those that stayed at home. If the alpha female has a litter, she may continue to go on hunts and one of the females of lower rank will tend the cubs in her absence. One of the most striking aspects of the integration is the hunt itself. In a pack they can attack and kill a large animal, such as a moose, but this requires coordinated tactics. They are capable of flanking maneuvers, of cutting individuals off from a herd, and then of systematically bringing down a large animal by biting from all sides until the animal is exhausted. Some of the flanking movements of wolves when attacking Dall sheep in Alaska, described by A. Murie (1944), are particularly impressive. The wolves will separate into two groups, one on the mountain side and the other hidden in the valley. The mountain group will stampede the sheep down into the valley where the other members of the pack are waiting, and then they all share the meal.

These details are interesting but they must not obscure the point I wish to make. It is simply that by behavior it has been possible to integrate groups of multicellular animals in ways that have become remarkably elaborate and sophisticated. These involve not only signals and the appropriate receptors between individuals, although these are essential, but they also include learning and memory, which come from experience. As is true in other animals, it is this accumulation of experience which makes alpha (and therefore breeding) individuals. Furthermore, there is the possibility that the brains of these social animals might invent new techniques of obtaining food, much the way Imo the Japanese macaque did; and these inventions, if they are effective, will certainly be passed on as memes to the other individuals of the group and probably to the subsequent generations. With behavioral plasticity the possibilities of integrating a social group in important and intricate ways become that much greater. To make a final leap to ourselves, we need think only of the myriad ways we make bonds between individual members of one group. Communication has now blossomed into language, and the subtlety of bond formation can be enormously complicated so that the bonds are not only different for different situations, but there is the possibility of a whole range of intensities of the bonds.

Somatic complexity level: Isolation. If we look at the smallest units, cells, we find good examples of isolating mechanisms. They, along with those examples from larger multicellular organisms, are instances where the isolation favors some somatic function, most often improving the efficiency of feeding. A particularly instructive case is negative chemotaxis in cellular slime mold amoebae.

As we have already discussed in some detail, the basis of signaling between cells involves a chemical process. There are signal molecules secreted by cells and there are receptor molecules, often in the membranes of cells, that combine with the signals. The receptors are large proteins, and when joined with the signal they are able to initiate a response reaction. Let us first look at this molecular signal-response system in single-cell organisms and see if we can imagine some of its early evolutionary history.

As mentioned earlier, some bacteria can orient in chemical gradients, for they have receptors which react with substances that might be found in their environment: food substances such as sugars or amino acids, or noxious substances of various sorts. This suggests that receptors arose first in evolution—not in response to signals given off by other cells of their own kind, but in response to environmental cues.

The next step is when cells synthesize both the receptor and the signal molecule; they become self-signaling. There is a good example of this among separate, feeding amoebae of cellular slime molds. They tend to repel one another so that they seem to be evenly spread as they feed, a phenomenon thought to provide optimal grazing when there is an even lawn of bacterial food. It can be clearly demonstrated that each cell gives off a chemical signal, and if there is a collection of cells in one spot they will slowly separate from one another by negative chemotaxis. Another good example of this is found in the development of the pigment cells in a species of salamander where these cells are evenly spaced over the body. There is excellent evidence that this spacing is also due to negative chemotaxis between the cells.

Another manifestation of isolation at the somatic diversity level is the formation of territories. This is found at different size levels, and a good example at the smallest level is found in the cellular slime molds. The area of amoebae that join one aggregate appears to be under more than one control mechanism, but the principle control is that once a center is formed it produces an inhibitor which prevents other centers from forming in its vicinity; it stakes out its territory by means of a chemical inhibitor (Fig. 83). Furthermore, as the fruiting bodies arise from the central cell masses they give off a repellent gas so that, in a dense culture, the spore masses end up to some degree equidistant from one another. In both these instances—the aggregation territories and the spacing of fruiting bodies—it is presumed that one likely selective advantage is for a more effective dispersal of spores. This is a reasonable hypothesis, but it would be valuable to test the idea experimentally.

In higher plants a similar kind of spacing is often found. Trees may space themselves in a forest, or plants in an arid area may be beautifully separated in a nonrandom fashion. The reasons for such spacing are varied: it may be competition for space in the sun (forest trees) or for water (desert plants). In some instances a plant will give off toxic substances which inhibit the growth of other plants in its vicinity; for instance, this is a strategy of walnut trees. Again in these cases the selective advantages are clear: it is the competition for resources that favors this isolation in the form of distancing.

There is a more general phenomenon found in many animals, characteristic of all groups of vertebrates: it is the formation of territories. This is essentially what the social insects are doing when they defend their nest site. A wolf pack will hunt in a specific area, and will mark off this area by periodically urinating along the boundary—a warning to other wolf packs that the territory has already been claimed as a hunting preserve. The same situation is found in many other mammals, although some have territories, or home ranges, that overlap to varying degrees.

Territories were initially described in birds and there are countless examples. Breeding pairs set up their nest in a desired area and then chase away any intruder from the whole area surrounding the nest; often the borders of this area are very sharply defined. Usually the male is the main guardian of the territory, and he uses his song as a warning to intruders to stay away. Without going into all the variations for birds, reptiles, amphibians, and fish, let me say that it is a very general phenomenon, and even found in varying degrees of clarity among insects and other arthropods. In all these instances it is thought that again the reason natural selection has produced such a system is to partition resources effectively.

Genomic complexity level: Integration. We are now at a level that is concerned with differences between species, and therefore integration must mean the coming together of more than one species to form an integrated unit. The formation of hybrids in nature could be considered a special case of such integration, but far more appropriate is the myriad of examples of symbiosis found throughout the animal and plant kingdoms. It is not always clear that there is a mutual benefit; the advantage may be one-sided. However, there is no doubt that there are advantages and that these advantages are sufficiently significant so that the associations are favored by natural selection. The main evidence for such a sweeping assertion comes from the extraordinary numbers that exist on earth today. To be convinced of their magnitude one need only consult the encyclopedic book of P. Buchner (1965) on microplant symbionts found in the cells of animals. The richness of his examples is quite remarkable and it covers only one specific kind of symbiosis. There are associations between microbes, and in fact such associations may have played a significant part in the evolution of eukaryotes (L. Margulis 1981). There are numerous symbioses between two species of plants such as lichens, and two species of animals such as the many parasitic associations between various worms and mammals, or between certain species of fish that live among the tentacles of cnidarians such as sea anemones.

We can immediately see from this fragmentary list that one species may be inside the cell of another (as in Buchner's endosymbioses), or there may be a collection of cells of two species (as in lichens), or one partner may be in the gut of the other (as in cellulose-digesting microorganisms inside mammals), or, finally, the partners may be quite separate from one another but have worked out a complex behavioral cooperation (such as the tiny fish that clean the skin of larger ones, thereby gaining protection and food in exchange for grooming). The degree of intimacy between the partners can vary enormously, but the evolutionary success of symbiosis cannot be questioned for it is so common and abundant a phenomenon.

Genomic complexity level: Isolation. The significance of isolation at this level is that without such isolation the mechanism for producing a controlled amount of variation would be impossible; every gain in any competitive advantage would be lost by immediate hybridization if there were no isolating mechanisms to prevent a mixing back of the genes which had successfully come to differ in sister genomes. Each successful genetic advance must be protected against being homogenized out of existence. To begin with primitive organisms, our first example will be from cellular slime molds.

There are a number of different species of cellular slime molds, and these are characterized by their ability to keep separate from one another to avoid making chimaeric fruiting bodies containing the cells of two species. This was first examined by K. B. Raper and C. Thom (1941), who discovered that two mechanisms were involved. They mixed different pairs of species, in each case marking the cells of one of the species so the amoebae could be distinguished. In one such mixture (*Dictyostelium discoideum* with *D. mucoroides*) the amoebae did co-aggregate, but at the center of the aggregation the cells of the two species sorted out and produced two adjacent fruiting bodies, each containing the spores of only one of the species (Fig. 85). In the other case (*D. discoideum* mixed with *Polysphondylium violaceum*) there was no co-aggregation; the cells went to separate centers, even though the aggregation streams overlapped (Fig. 83). We now have a molecular understanding of both these cases, although it is still relatively crude in the first instance. There the motile cells of the two species differ in their adhesive properties and as a result they probably sort out by differential adhesion. What is not known are the molecular details of the cell surface differences of different species. In the second case the difference between the two species is that they have different signal-receptor systems. Here the chemical indentity of the attractants (or acrasins) is known: one is cyclic AMP and the other is a dipeptide called glorin.

The latter case is the one of central interest to us here. In these two species (*D. discoideum* and *P. violaceum*) not only is the acrasin of each one a radically different chemical, but in order for the system to work there must be different protein receptors at the cell surface, as well as different inactivating enzymes or acrasinases. It is difficult to estimate how many different basic proteins (and their genes) might be involved for each acrasin, but four would be a minimal and most conservative estimate.

If we now look to other species of dictyostelids, we see further evidence for acrasin diversity. It has been known for some time that most

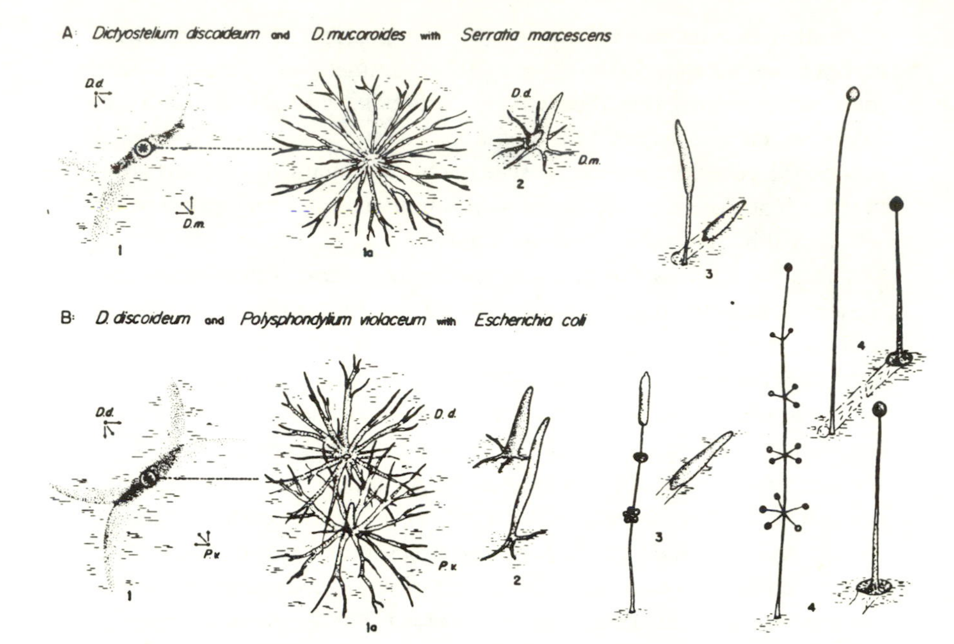

Fig. 85. Diagrams to show how different combinations of species of cellular slime molds avoid forming a chimaeric fruiting body. *A*, two species that coaggregate but their tips subsequently separate at the center of the aggregate. *B*, two species that have different chemotactic substances and therefore form separate aggregations. (From Raper and Thom 1941.)

dictyostelids also respond chemotactically to gradients of folic acid and pterin derivatives. There is some evidence that for many species this response is especially strong during the feeding or vegetative stage, and therefore perhaps involved in finding food, for bacteria exude folic acid. Recently T. M. Konijn and his colleagues in Leiden have shown that two other species are known to use different folic acid derivatives for their respective acrasins.

This means that we have a good idea of the identity of four different acrasins, and some evidence for the existence of four others. Undoubtedly more exist and will be discovered when all the fifty dictyostelid species have been analyzed. In other words, the evolution of new signal-receptor systems has been a major factor in producing and maintaining diversity in the cellular slime molds.

If this is so, how do new aggregation chemotaxis systems arise? Since at least four and probably even more specific proteins are involved, it is inconceivable that an equivalent number of new genes could have simultaneously appeared by new gene mutation. A more reasonable hypothesis is that cellular slime mold amoebae have more than one signal-receptor system and just one of these is used in aggregation. Consider the case of *D. discoideum*, where the cells have folate receptors and a folic acid deaminase that removes extracellular folate. This system is not used for aggregation, but, as mentioned earlier, possibly for food seeking. P. N. Devreotes (1983) suggests an additional function: that folic acid might also potentiate, that is, enhance the sensitivity to acrasin.

In this case of slime mold cell isolation there is a direct connection with species formation. One might well ask why the amoebae bother to become different and isolated. Why do they not all remain capable of forever fusing with one another, so that the fruiting bodies become an increasing mixture of cells of diverse genetic constitutions? As L. W. Buss (1982, 1983) would argue, the answer undoubtedly lies in the fact that all cells are in competition, and if a genetic difference appears within a population of cells, and it gives those cells with the new mutation some advantage, they will try to avoid aggregation with the other cell types, so that they can selfishly exploit their newly acquired advantage. There is some evidence to support this from the work of E. G. Horn (1971), who showed that in a plot of ground four species of dictyostelids coexisted; if he isolated the bacteria from the same plot and tested in different ways each slime mold with the various strains of bacteria, there was a clear indication that each species of slime mold had preferences for different strains of bacteria. In other words, each species of slime mold has become specialized for a particular food niche, and they retain their individuality by avoiding the formation of chimaeras. Stated in terms of selection, there is strong positive selection pressure for fruiting bodies to avoid being chimaeric, because the presence of foreign genotypes could sap the reproductive success of the parental genotype, largely by reducing the number of its spores. Therefore, favorable genotypes rapidly acquire isolation mechanisms, for without them these genotypes would soon be wiped out. All new, effective genetic changes must be accompanied by further ones that favor isolation. Only in this way can they have a good chance of remaining fixed in the cell population.

If we now look at the cell activities within multicellular organisms that arise from a fertilized egg, we see no evidence of the isolation of the different cell types within the growing cell mass; all the cells seem to be harmoniously integrated. There are, however, two ways in which isolation mechanisms appear in this kind of organism.

One is a self-recognition system which is characteristic of vertebrates. They have a method for rejecting foreign cells which is based on a complex set of genetic instructions to cell surfaces, so that foreign cells with different proteins on their surface are destroyed. This is beautifully illustrated in the grafting of skin in mammals; only skin from genetically similar individuals will remain; all others will be sloughed off. This immunological rejection does not occur at all stages of development, but appears only late, near birth and thereafter. It is possible to make artificial chimaeras in early embryos, but normally this is not a problem, for the embryo is protected in an egg or *in utero* and foreign cells cannot reach it. Among invertebrates, which have not developed such an immunological mechanism, there is a self/nonself-recognition mechanism that resembles the one in slime molds. For instance, in colonial hydroids such as *Hydractinia* the stolons can fuse or avoid fusion, depending on their genetic constitution. Such isolation mechanisms are absent in higher plants, but that would be expected because of their manner of growth with meristems and hard cell walls, the kind of structure ill-suited to cell invasion because the cells themselves are not motile. It is known, however, that plants have surface mechanisms involving the production of specific substances that resist the invasive growth of pathogenic fungi.

The other way in which cellular isolation mechanisms play an important role is in fertilization. The fertilization reaction for all organisms is sufficiently specific so that it will permit the fusion of the male and female gamete, but only those of the same or closely related species. The mating reaction is either encouraged or, if strange species are involved, blocked by surface molecules present in the membranes of the gametes of both sexes. Specific, discriminatory molecules at the surface are found in all eukaryotic sexual organisms, from the most primitive protozoa and algae to higher plants and animals. By the general blocking of hybridization such fertilization specificity serves to keep species distinct and isolated. At first a population that has changed genetically may have no such barrier at fertilization with its parent population; but (as with slime molds) if the genetic difference is to be maintained between the two populations, it is inevitable that selection will also favor an isolating mechanism, and one of the most effective is a barrier at fertilization.

One interesting aspect of sexuality should be pointed out here. While it is an effective method to isolate genetically different populations, it is also clearly a method of integration. It brings together the genetic complement of two individuals and fuses them into one cell, with the consequence that they go on to produce a single, multicellular individual possessing the genes of both parents. So here, as in slime mold aggregation, we see the complementary nature of integration and isolation in one process.

Finally, let me discuss what is undoubtedly the most important aspect of isolation, that is, its role in controlling the amount of variation. This is seen in a straightforward manner if we look at the way in which species are formed, where, by isolation, two species arise from one.

Isolation of separate, incipient species populations arises by chance, not by a "selection for isolation." The most commonly supposed method is by geographic isolation. Individuals from the mainland population of some bird occupy an island and from that moment onward there is no further exchange with the mainland. Either by random drift or selection the island population eventually shows numerous phenotypic differences (and the corresponding gene differences) from the mainland population. After enough time has elapsed, if they are reunited, they are no longer capable of hybridization. There are many possible reasons. As pointed out above, it might be due to differences arising between the signal-response system of egg and sperm. It might be because the island differed in climate so that their mating seasons no longer correspond. Or it may be that in their courtship behavior certain differences have appeared in the pattern of the ritual that is off-putting to successful courtship. It could even be a slight difference in color, as N. G. Smith (1966) showed for gulls, where mating success between two species depends upon the color band surrounding the eye, one species being blue and the other yellow. By artificially painting these eye rings he could make them mate successfully. In all these cases the isolating mechanism appears to have arisen quite incidentally; it was probably not the immediate object of selection.

Consider now the case where two species arise in the same region, the so-called sympatric speciation. This is what is supposed to have happened to some of Darwin's finches in the Galapagos islands. There is good reason to believe that the islands were first colonized by a few finches of one species from the mainland. They have now turned into a dozen species. In some cases these new species no doubt evolved because they inhabited separate islands of the archipelago, but in others the species could have formed sympatrically. Let us assume that some of the seed-eating forms of these finches initially had a beak size that could crush two kinds of seed that were at the extreme size range of the beak's ability, that is, one seed was too small for effective harvesting, while the other was too big (Fig. 86). This would result in a bimodal selection: for birds with either larger beaks or smaller beaks, and the intermediate ones would be at a disadvantage. At first the population continues to be able to interbreed freely; but if hybrids are formed between a large-beak and a small-beak parent, they will be intermediate (making certain assumptions on how beak size is inherited) and assume there will be a strong selection pressure against hybrids. This will mean that those birds that pick mates with the same beak size, either large or small, will be favored and it can easily be imagined that some barrier,

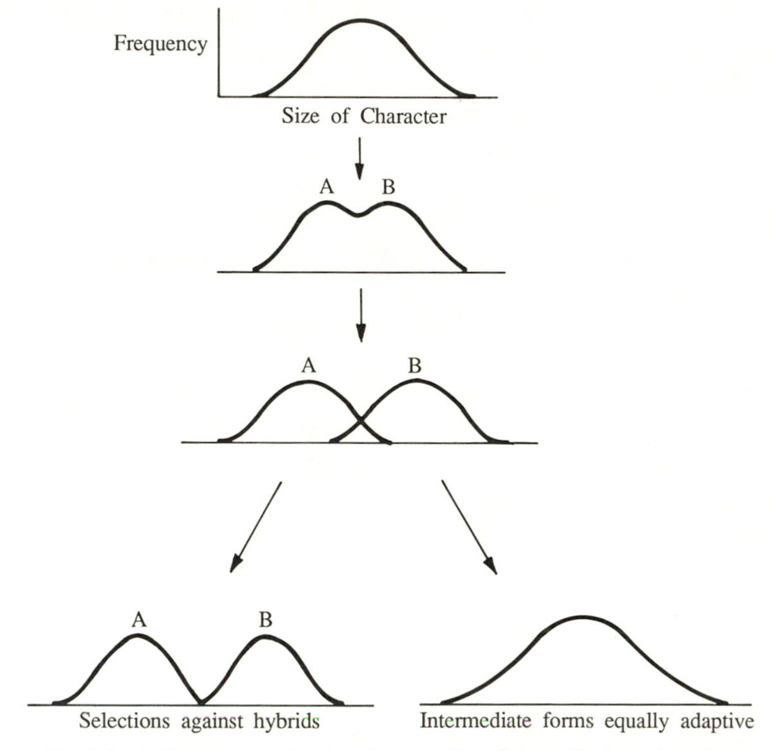

Fig. 86. A diagram showing how intermediate forms, if selected against, can lead to the isolation of two separate forms that ultimately may become new species.

either behavioral or even involving the chemistry of fertilization, if it arose, would be immediately favored by selection. In this case too, the initial isolating mechanism is a selection for something else—seed size causing optimal beak sizes—and this is followed by an isolation mechanism.

Sexuality is a system of inheritance that has been enormously successful, for we find it present in almost all organisms. It has meant reproductive success and is responsible for major evolutionary progress. One of the conditions necessary for the sexual genetic system to operate is to have mechanisms of isolation. By controlling the exchange of genes between a mating pair it is possible to produce sufficiently variable offspring, so that some will be suited for the same environment and others for a new, unexpected environment. (If the environment is constant, then the most efficient strategy is asexual reproduction which repeats the same successful genetic constitution each generation.) By keeping control of the amount of genetic diversity in a population, one has the optimum adaptability for a variety of environmental conditions. In other words, the amount of genetic diversity is in itself adaptive, and this can be superbly controlled by the sexual system, and one of the needed mechanisms for that control is isolation.

SUMMARY

These three insights help us to see into the evolution of complexity. The first insight shows that there are two ways in which complexity is controlled by the genome: one is a direct control such as one finds in the differences between species, and the other is a distant genetic control, for the differences which divide the labor are genetically identical or nearly so; the diversity is somatic such as one finds within a species. As evolution proceeded on the surface of the earth, there has been a progressive increase in size and complexity, and the second insight shows that for any one size group of organisms there is an inverse relation between the somatic complexity (internal division of labor) and the genetic complexity (species diversity), that is, the larger and more complex a type of organism, the fewer species of that type will exist in any one environment and vice versa. Finally, the third insight shows that through signal-response systems there has been a concurrent integration and isolation of living units at both the levels of somatic and genomic complexity. There are strong arguments in favor of the idea that integration and isolation are simultaneously under positive selection. Thus we see that the genetic mechanism, which is selected for and is the basis of natural selection, has been able to produce and control complexity at two different levels (somatic and genomic), and that this complexity is correlated and constrained by size. One of the ways of allowing evolutionary change to occur is by having a selection for an appropriate balance of both the integration and the isolation of living units. In this way reproduction and the control of variation, the essential components of natural selection which have led to the evolution of complexity can occur.

Further References

For current reviews of cellular slime mold aggregation, see P. C. Newell (1981) and P. N. Devreotes (1982). For discussions of control mechanisms of spore-stalk proportions, see H. K. MacWilliams and J. T. Bonner (1979), J. H. Morrissey (1982), and C. J. Weijer and A. J. Durston (1985). For a recent summary of wolf behavior, see L. D. Mech (1970). Negative chemotaxis in the amoebae of cellular slime molds is discussed in M. T. Keating and J. T. Bonner (1977), and P.I.J. Kakabeeke et al. (1979). A review of the spacing of pigment cells in amphibians will be found in V. C. Twitty (1949). Territories and zones of inhibition of rival centers are described in B. M. Shaffer (1963), J. T. Bonner and M. E. Hoffman (1963), and D. R. Waddell (1982). There is also a spacing substance for rising spore masses (J. T. Bonner and M. R. Dodd 1962). Differential cell adhesion in slime molds is examined by A. Nicol and D. R. Garrod (1978), and J. Sternfeld (1979). Recent work on different chemical aggregation signals in slime molds is reviewed in Bonner (1982b, 1983). The early work on folic acid as a chemoattractant in slime molds is found in P. Pan et al. (1972, 1975), and a more recent review is R.J.W. De Wit and T. M. Konijn (1983).

Bibliography

Aiken, D. E., and F. E. Barton II. 1983. Rumen microbial attachment and degradation of plant cell walls. *Federation Proceedings* 42:114–121.

Alberts, B., D. Bray, J. Lewis, M. Raff, K. Roberts, and J. D. Watson. 1983. *Molecular Biology of the Cell*. Garland Publishing, New York.

Alexander, R. D. 1979. *Darwinism and Human Affairs*. University of Washington Press, Seattle.

Alexander, R. D., J. L. Hoogland, R. D. Howard, K. M. Noonan, and P. W. Sherman. 1979. Sexual dimorphisms and breeding systems in pinnipeds, ungulates, primates and humans. In *Evolutionary Biology and Human Behavior*, pp. 402–435. N. A. Chagnon and W. Irons, eds. Duxbury Press, North Scituate, Mass.

Bonner, J. T. 1952. *Morphogenesis*. Princeton University Press, Princeton, N.J.

Bonner, J. T. 1958. *The Evolution of Development*. Cambridge University Press, New York.

Bonner, J. T. 1965. *Size and Cycle*. Princeton University Press, Princeton, N.J.

Bonner, J. T. 1974. *On Development*. Harvard University Press, Cambridge, Mass.

Bonner, J. T. 1980. *The Evolution of Culture in Animals*. Princeton University Press, Princeton, N.J.

Bonner, J. T., ed. 1982a. *Evolution and Development*. Dahlem Conference. Springer-Verlag, Berlin.

Bonner, J. T. 1982b. Evolutionary strategies and development constraints in the cellular slime molds. *Amer. Nat*. 119:530–552.

Bonner, J. T. 1983. Chemical signals of social amoebae. *Sci. Amer.* (March):106–112.

Bonner, J. T., and M. R. Dodd. 1962. Evidence for gas-induced orientation in the cellular slime molds. *Develop. Biol*. 5:344–361.

Bonner, J. T., and M. E. Hoffman. 1963. Evidence for a substance responsible for the spacing pattern of aggregation and fruiting in the cellular slime molds. *J. Embryol. Exp. Morph*. 11:571–589.

Bradshaw, A. D. 1965. Evolutionary significance of phenotype plasticity in plants. *Adv. Genet*. 13:115–155.

Britten, R. J., and E. H. Davidson. 1969. Gene regulation for higher cells: A theory. *Science* 165:349–358.

Brown, J. H., and B. A. Maurer. 1986. Body size, ecological dominance and Cope's rule. *Nature* 324:248–250.

Buchner, P. 1965. *Endosymbiosis of Animals with Plant Microorganisms*. Wiley, New York.

Buss, L. W. 1982. Somatic cell parisitism and the evolution of somatic tissue compatibility. *Proc. Nat. Acad. Sci. USA* 79:5337–5341.

Buss, L. W. 1983. Evolution, development, and the units of selection. *Proc. Nat. Acad. Sci. USA* 80:1387–1391.

Buss, L. W. 1987. *The Evolution of Individuality*. Princeton University Press, Princeton, N.J.

Calder, W. A., III. 1984. *Size, Function, and Life History*. Harvard University Press, Cambridge, Mass.

Campbell, R. D., R. K. Josephson, W. E. Schwab, and N. B. Rushforth. 1976. Excitability of nerve-free hydra. *Nature* 262:388–390.

Carlile, M. J. 1980. From prokaryote to eukaryote: Gains and losses. *Symp. Soc. Gen. Microbiol.* 30:1–40.

Cavalli-Sforza, L. L., and M. W. Feldman. 1978. Towards a theory of cultural evolution. *Interdisciplinary Sci. Rev.* 3:99–107.

Coates, A. G., and J.B.C. Jackson. 1985. Morphological themes in the evolution of clonal and aclonal marine invertebrates. In *Population Biology and Evolution in Clonal Organisms*, pp. 67–106. J.B.C. Jackson, L. W. Buss, and R. E. Cook, eds. Yale University Press, New Haven, Conn.

Cole, B. J. 1981. Dominance hierarchies in *Leptothorax* ants. *Science* 212:83–84.

Cowan, W. M., J. W. Fawcett, D.D.M. O'Leary, and B. B. Stanfield. 1984. Regressive events in neurogenesis. *Science* 225:1258–1265.

Cox, P. A., and J. A. Sethian. 1985. Gamete motion, search, and the evolution of anisogamy, oogamy, and chemotaxis. *Amer. Nat.* 125:74–101.

Curio, E., V. Ernst, and W. Vieth. 1978. Cultural transmission of enemy recognition: One function of mobbing. *Science* 202:899–901.

Dalcq, A. M. 1938. *Form and Causality in Development*. Cambridge University Press, New York.

Darwin, C. 1859. *On the Origin of Species*. Reprint edition, Harvard University Press, Cambridge, Mass.

Darwin, C. 1871. *Descent of Man*. Reprint edition, Princeton University Press, Princeton, N.J.

Dawkins, R. 1976. *The Selfish Gene*. Oxford University Press, New York.

Dawkins, R. 1982. *The Extended Phenotype*. W. H. Freeman, San Francisco.

Dawkins, R., and J. R. Krebs. 1979. Arms races between and within species. *Proc. Roy. Soc. London B*. 205:489–511.

de Beer, G. R. 1958. *Embryos and Ancestors*. Clarendon Press, Oxford. (This book first appeared in 1930 in an earlier form called *Embryology and Evolution*.)

Devreotes, P. N. 1982. Chemotaxis. In *The Development of* Dictyostelium discoideum. W. F. Loomis, ed. Academic Press, New York.

Devreotes, P. N. 1983. The effect of folic acid on cAMP-elicited cAMP production in *Dictyostelium discoideum*. *Develop. Biol.* 95:154–162.

De Wit, R.J.W., and T. M. Konijn. 1983. Pterins and folates as extracellular signals for chemotaxis and differentiation in the cellular slime molds. In *Biochemical and Clinical Aspects of Pteridines*, vol. 2, pp. 383–400. H. C. Curtius, W. Pfleiderer, and H. Wachter, eds. Walter de Gruyer & Co., Berlin.

Diamond, J. M. 1973. Distributional ecology of New Guinea birds. *Science* 179:759–769.

Duboscq, O., and O. Tuzet. 1935. L'ovogénèse, la fécondation et les premiers stades du développement des éponges calcaires. *Arch. Zool. Expér. et Gén.* 79:157–316.

Durham, W. H. 1982. Interactions of genetic and cultural evolution: Models and examples. *Human Ecology* 10:289–323.

Elton, C. S. 1927. *Animal Ecology*. Macmillan, New York.

Fankhauser, G. 1955. The role of nucleus and cytoplasm. In *Analysis of Development*, pp. 126–150. B. H. Willier, P. A. Weiss, and V. Hamburger, eds. W. B. Sanders, Philadelphia.

Filosa, M. F. 1962. Heterocytosis in cellular slime molds. *Amer. Nat.* 96:79–91.

Fisher, R. A. 1930. *The Genetical Theory of Natural Selection*. Clarendon Press, Oxford.

Friedmann, H., J. Kern, and J. H. Hurst. 1957. The domestic chick: A substitute for the honey-guide as a symbiont with cerolytic microorganisms. *Amer. Nat.* 91:321–325.

Frye, T. C., G. B. Rigg, and W. C. Crandall. 1915. The size of kelps on the Pacific coast of North America. *Bot. Gaz.* 60:473–482.

Garstang, W. 1922. The theory of recapitulation: A critical restatement of the biogenetic law. *J. Linn. Soc. Zool.* 35:81–101.

Ghiselin, M. T. 1974. *The Economy of Nature and the Evolution of Sex*. Univ. of California Press, Berkeley.

Gierer, A. 1974. Hydra as a model for the development of biological form. *Sci. Amer.* 231 (December):44–54.

Gingerich, P. D. 1974. Stratigraphic record of early Eocene *Hyposadus* and the geometry of mammalian phylogeny. *Nature* 284:107–109.

Gingerich, P. D. 1983. Rates of evolution: Effects of time and temporal scaling. *Science* 222:159–161.

Goldschmidt, R. B. 1940. *The Material Basis of Evolution*. Yale University Press, New Haven, Conn.

Gould, J. L. 1982. *Ethology: The Mechanism of Evolution and Behavior*. W. W. Norton & Co., New York.

Gould, S. J. 1977. *Ontogeny and Phylogeny*. Harvard University Press, Cambridge, Mass.

Gould, S. J., and N. Eldredge. 1977. Punctuated equilibria: The tempo and mode of evolution reconsidered. *Paleobiology* 3:115–151.

Greenspan, R. J. 1988. Genes as bits for nervous system development. In *Advances in Cognitive Science*, pp. 114–127. M. Kochen and H. Hastings, eds. Westview Press, Boulder, Colorado.

Griffin, D. R. 1984. *Animal Thinking*. Harvard University Press, Cambridge, Mass.

Hamburger, V. 1980. Embryology and the modern synthesis in evolutionary theory. In *The Evolutionary Synthesis: Perspectives on the Unification of Biology*, pp. 97–112. E. Mayr and W. B. Provine, eds. Harvard University Press, Cambridge, Mass.

Harper, J. L. 1977. *Population Biology of Plants*. Academic Press, New York.

Harvey, P. H., and P. M. Bennett. 1983. Brain size, energetics, ecology and life history patterns. *Nature* 306:314–316.

Hofmann, H. J., and J. W. Schopf. 1983. Early proterozoic microfossils. In *Earth's*

Earliest Biosphere: Its Origin and Evolution, pp. 321–360. J. W. Schopf, ed. Princeton University Press, Princeton, N.J.

Horn, E. G. 1971. Food competition among cellular slime molds. *Ecology* 52:475–484.

Horn, H. S., and D. I. Rubenstein. 1984. Behavioral adaptations and life history. In *Behavioral Ecology: An Evolutionary Approach*, pp. 279–298. J. R. Krebs and N. B. Davies, eds. Blackwell Scientific Publications, Oxford.

Hutchinson, G. E. 1959. Homage to Santa Rosalia, or why are there so many kinds of animals. *Amer. Nat.* 93:145–159.

Hutchinson, G. E., and R. H. MacArthur. 1959. A theoretical ecological model of size distributions among species of animals. *Amer. Nat.* 93:117–125.

Hyman, L. H. 1940. *The Invertebrates: Protozoa through Ctenophora*. McGraw-Hill, New York.

Hyman, L. H. 1942. The transition from the unicellular to the multicellular individual. *Biol. Symposia* 8:27–42.

Jackson, J.B.C., L. W. Buss, and R. E. Cook, eds. 1985. *Population Biology and Evolution of Clonal Organisms*. Yale University Press, New Haven, Conn.

James, F. C. 1970. Geographic variation in birds and its relationship to climate. *Ecology* 51:365–390.

Jarvis, J.U.M. 1981. Eusociality in a mammal: Cooperative breeding in naked mole-rat colonies. *Science* 212:571–573.

Jenkins, P. F. 1978. Cultural transmission of song patterns and dialect development in a free-living bird population. *Anim. Behav.* 25:50–78.

Jennings, H. S. 1905. *Behavior of the Lower Organisms*. Reprint edition, Indiana University Press, Bloomington.

Kakabeeke, P.I.J., R.J.W. De Wit, S. P. Kohtz, and T. M. Konijn. 1979. Negative chemotaxis in *Dictyostelium* and *Polysphondylium*. *Exp. Cell. Res.* 124:429–433.

Kandel, E. R. 1976. *Cellular Basis of Behavior*. W. H. Freeman, San Francisco.

Katz, M. J. 1982. Ontogenetic mechanisms: The middle ground of evolution. In *Evolution and Development*. J. T. Bonner, ed. Springer-Verlag, Berlin.

Katz, M. J. 1983. Ontophyletics: Studying evolution beyond the genome. *Perspect. in Bio. and Med.* 26:323–333.

Katz, M. J. 1986. Is evolution random? In *Development as an Evolutionary Process*, pp. 285–315. R. A. and E. C. Raff, eds. MBL Lectures in Biology. Alan R. Liss, New York.

Kaufmann, S. A. 1969. Metabolic stability and epigenesis in randomly constructed genetic nets. *J. Theoret. Biol.* 22:437–467.

Keating, M. T., and J. T. Bonner. 1977. Negative chemotaxis in cellular slime molds. *J. Bact.* 130:144–147.

Kelland, J. L. 1977. Inversion in *Volvox* (Chlorophyceae). *J. Phycol.* 13:373–378.

Kemp, P., and M. D. Bertness. 1984. Snail shape and growth rates: Evidence for plastic shell allometry. *Proc. Nat. Acad. Sci. USA*. 81:811–813.

King, A. P., and M. J. West. 1977. Species identification in the North American cowbird: Appropriate responses to abnormal song. *Science* 195:1002–1004.

King, M.-C., and A. C. Wilson. 1975. Evolution at two levels in humans and chimpanzees. *Science* 188:107–116.

Kirschner, M., J. Newport, and J. Gerhart. The timing of early developmental events in *Xenopus*. 1985. *Trends in Genetics* 1 (February):41–47.

Kochert, G. 1975. Developmental mechanisms in *Volvox* reproduction. In *Developmental Biology of Reproduction*, pp. 55–90. C. C. Markert and J. Papaconstantinou, eds. 33rd Symposium, Soc. Dev. Biol. Academic Press, New York.

Konishi, M. 1965. The role of auditory feedback in the control of vocalizations in the white-crowned sparrow. *Zeit. Tierpsychol.* 22:770–783.

Kuhn, A. 1971. *Lectures on Developmental Physiology*. Springer-Verlag, New York.

Kurtén, B. 1959. Rates of evolution in fossil mammals. *Cold Spring Harbor Symposium: Quant. Biol.* 24:205–215.

Kurtén, B. 1968. *Pleistocene Mammals of Europe*. Aldine, Chicago.

Lack, D. 1947. *Darwin's Finches*. Cambridge University Press, New York.

Levinton, J. S. 1986. Developmental constraints and evolutionary saltations: A discussion and critique. In *Genetics, Development and Evolution*, pp. 253–288. J. P. Gustafson, G. L. Stebbins, and F. J. Ayala, eds. 17th Stadler Genetics Symposium, University of Missouri. Plenum, New York.

Lomolino, M. V. 1985. Body size of mammals on islands: The island rule reexamined. *Amer. Nat.* 125:310–316.

Lumsden, C. J., and Wilson, E. O. 1981. *Genes, Mind and Culture: The Coevolutionary Process*. Harvard University Press, Cambridge, Mass.

MacArthur, R. H., and E. O. Wilson. 1967. *Island Biogeography*. Princeton University Press, Princeton, N.J.

McMahon, T. A. 1973. Size and Shape in Biology. *Science* 179:1201–1204.

McMahon, T. A., and J. T. Bonner. 1983. *Size and Life*. Scientific American Library: W. H. Freeman & Co.

MacWilliams, H. K., and J. T. Bonner. 1979. The prestalk-prespore pattern in cellular slime molds. *Differentiation* 12:1–22.

Margulis, L. 1981. *Symbioses in Cell Evolution*. W. H. Freeman, San Francisco.

May, R. M. 1978. The dynamics and diversity of insect faunas. In *Diversity of Insect Faunas*. Symp. Roy. Entomol. Soc. London No. 9, pp. 188–204. L. A. Mound and N. Waloff, eds. Blackwell Scientific Publications.

Mech, L. D. 1970. *The Wolf: The Ecology and Behavior of an Endangered Species*. Natural History Press, New York.

Meinhardt, H. 1982. *Models of Biological Pattern Formation*. Academic Press, New York.

Morrissey, J. H. 1982. Cell proportioning and pattern formation. In *Development of* Dictyostelium discoideum, pp. 411–449. W. F. Loomis, ed. Academic Press, New York.

Morse, D. R., J. H. Lawton, and M. M. Dodson. 1985. Fractal dimension of vegetation and the distribution of arthropod body lengths. *Nature* 314:731–733.

Murie, A. 1944. *The Wolves of Mount McKinley*. Fauna of the National Parks of the United States, Fauna Series No. 5. U. S. Department of the Interior, Washington, D.C.

Newell, N. D. 1949. Phyletic size increase, an important trend illustrated by fossil invertebrates. *Evolution* 3:103–124.

Newell, P. C. 1981. Chemotaxis in the cellular slime moulds. In *Biology of the Chemotactic Response*. J. M. Lackie and P. C. Wilkinson, eds. *Soc. Exp. Biol. Seminar Series* 12:89–114.

Nicol, A., and D. R. Garrod. 1978. Mutual cohesion and cell sorting-out among four species of cellular slime moulds. *J. Cell. Sci.* 32:377–387.

Oster, G. F., J. O. Murray, and A. K. Harris. 1983. Mechanical aspects of mesenchymal morphogenesis. *J. Embryol. Exp. Morph.* 78:83–125.

Oster, G. F., and E. O. Wilson. 1978. *Caste and Ecology in the Social Insects*. Princeton University Press, Princeton, N.J.

Pan, P., E. M. Hall, and J. T. Bonner. 1972. Folic acid as a second chemotactic substance in the cellular slime moulds. *Nature New Biol.* 237:181–182.

Pan, P., E. M. Hall, and J. T. Bonner. 1975. Determination of the active portion of the folic acid molecule in cellular slime mold chemotaxis. *J. Bact.* 122:185–191.

Peters, R. H. 1983. *The Ecological Implications of Size*. Cambridge University Press, New York.

Peterson, R. O., R. E. Page, and K. M. Dodge. 1984. Wolves, moose, and the allometry of population cycles. *Science* 224:1350–1352.

Pianka, E. R. 1983. *Evolutionary Ecology*. Harper & Row, New York.

Picken, L.E.R. 1960. *The Organization of Cells and Other Organisms*. Oxford University Press, New York.

Poirier, F. E. 1977. Introduction in *Primate Bio-Social Development*, pp. 1–39. S. Chevalier-Skolnikoff and F. E. Poirier, eds. Garland Publishing, Inc., New York.

Raff, R. A., and T. C. Kaufman. 1983. *Embryos, Genes, and Evolution*. Macmillan, New York.

Raper, K. B., and C. Thom. 1941. Interspecific mixtures in the Dictyosteliaceae. *Amer. J. Bot.* 28:69–78.

Richerson, P. J., and R. Boyd. 1985. *Culture and the Evolutionary Process*. University of Chicago Press, Chicago.

Rubenstein, D. I. 1982. Reproductive value and behavioral strategies: Coming of age in monkeys and horses. *Perspectives in Ethology* 5:469–487.

Schluter, D., T. D. Price, and P. R. Grant. 1985. Ecological character displacement in Darwin's finches. *Science* 227:1056–1059.

Schmalhausen, I. I. 1949. *Factors of Evolution*. Blakiston, Philadelphia.

Schmidt-Nielsen, K. 1984. *Scaling: Why is Animal Size So Important*? Cambridge University Press, New York.

Schneirla, T. C. 1953. Modifiability in insect behavior. In *Insect Physiology*, pp. 723–747. John Wiley, New York.

Scholander, P. F. 1955. Evolution of climatic adaptation in homeotherms. *Evolution* 9:15–26.

Seyfarth, R. M., and D. L. Cheney. 1984. The natural vocalizations of nonhuman primates. *Trends in Neurosciences* 7 (March):66–73.

Shaffer, B. M. 1963. Inhibition by existing aggregations of founder differentiation in the cellular slime mould *Polysphondylium violaceum*. *Exp. Cell. Res.* 31:432–435.

Silander, J. A., Jr. 1985. Microevolution in clonal plants. In *Population Biology and Evolution in Clonal Organisms*, pp. 107–152. J.B.C. Jackson, L. W. Buss, and R. E. Cook, eds. Yale University Press, New Haven, Conn.

Simon, H. A. 1962. The architecture of complexity. *Proc. of the Amer. Phil. Soc.* 106:467–482.

Smith, N. G. 1966. Evolution of some arctic gulls (*Larus*): An experimental study of isolating mechanisms. *Ornithol. Monogr.* 4:1–99.

Smith, W. J. 1977. *The Behavior of Communicating*. Harvard University Press, Cambridge, Mass.

Solbrig, O. T., and D. J. Solbrig. 1984. Size inequalities and fitness in plant populations. *Oxford Surveys in Evol. Biol.* 1:139–157.

Sparrow, A. H., H. J. Price, and A. G. Underbrink. 1972. A survey of DNA content per cell and per chromosome of prokaryotic and eukaryotic organisms: Some evolutionary considerations. *Brookhaven Symp. Biol.* 23:451–494.

Stanley, S. M. 1973. An explanation of Cope's rule. *Evolution* 27:1–26.

Stanley, S. M. 1979. *Macroevolution*. W. H. Freeman and Co., San Francisco.

Stebbins, L. G., Jr. 1950. *Variation and Evolution in Plants*. Columbia University Press, New York.

Stent, G. S. 1985. Thinking in one dimension: The impact of molecular biology on development. *Cell* 40:1–2.

Sternfeld, J. 1979. Evidence for differential cellular adhesion vs. the mechanism of sorting-out of various cellular slime mold species. *J. Embryol. Exp. Morph.* 53:163–178.

Sugiyama, T., and T. Fujisawa. 1978. Genetic analysis of developmental mechanisms in Hydra. II. Isolation and characterization of an interstitial cell-deficient strain. *J. Cell Sci.* 29:35–52.

Thompson, D'Arcy W. 1942. *On Growth and Form*. Cambridge University Press, New York.

Tiffney, B. H., and K. J. Niklas. 1985. Clonal growth in land plants: A paleobotanical perspective. In *Population Biology and Evolution in Clonal Organisms*, pp. 35–66. J.B.C. Jackson, L. W. Buss, and R. E. Cook, eds. Yale University Press, New Haven, Conn.

Tilman, D. 1982. *Resource Competition and Community Structure*. Princeton University Press, Princeton, N.J.

Turing, A. M. 1952. The chemical basis of morphogenesis. *Phil. Trans. Roy. Soc. London B*. 237:37–72.

Twitty, V. C. 1949. Developmental analysis of amphibian pigmentation. *Growth, Symposium* 9:133–161.

Viamontes, G., L. Fochtmann, and D. Kirk. 1979. Morphogenesis in *Volvox*: Analysis of critical variables. *Cell* 17:537–550.

Waddell, D. R. 1982. The spatial pattern of aggregation centres in the cellular slime mold. *J. Embryol. Exp. Morph.* 70:75–98.

Waddington, C. H. 1940. *Organizers and Genes.* Cambridge University Press, New York.

Waddington, C. H. 1957. *The Strategy of the Genes.* Allen and Unwin, London.

Wainwright, S. A., W. D. Biggs, J. D. Currey, and J. M. Gosline. 1976. *Mechanical Design in Organisms.* E. Arnold, London.

Weijer, C. J., and A. J. Durston. 1985. Influence of cyclic AMP and hydrolysis products on cell type regulation in *Dictyostelium discoideum. J. Embryol. Exp. Morph.* 86:19–37.

Wheeler, W. M. 1911. The ant colony as an organism. *J. Morph.* 22:307–325.

Whyte, L. L. 1965. *Internal Factors in Evolution.* George Braziller, New York.

Williams, G. C. 1966. *Adaptation and Natural Selection: A Critique of Some Current Evolutionary Thought.* Princeton University Press, Princeton, N.J.

Wilson, E. O. 1971. *The Insect Societies.* Harvard University Press, Cambridge, Mass.

Wilson, E. O. 1975. *Sociobiology: The New Synthesis.* Belknap Press of Harvard University Press, Cambridge, Mass.

Wolk, C. P. 1982. Heterocysts. In *The Biology of Cyanobacteria.* N. G. Carr and B. A. Whitton, eds. Blackwell Scientific Publications, Oxford.

Index